EMULSIFYING AGENTS

EMULSIFYING AGENTS

An Industrial Guide

by

Ernest W. Flick

np NOYES PUBLICATIONS
Park Ridge, New Jersey, U.S.A.

Copyright © 1990 by Ernest W. Flick
No part of this book may be reproduced or utilized in any form or by any means, electronic or mechanical, including photocopying, recording or by any information storage and retrieval system, without permission in writing from the Publisher.
Library of Congress Catalog Card Number: 89-70980
ISBN: 0-8155-1225-2
Printed in the United States

Published in the United States of America by
Noyes Publications
Mill Road, Park Ridge, New Jersey 07656

10 9 8 7 6 5 4 3 2 1

Library of Congress Cataloging-in-Publication Data

Flick, Ernest W.
  Emulsifying agents : an industrial guide / by Ernest W. Flick.
      p.   cm.
  ISBN 0-8155-1225-2 :
  1. Emulsions. I. Title.
  TP156.E6F5   1990
  660'.294514--dc20   89-70980
                                                       CIP

To
The Present Hasty Generation
Glenice, Colby, Clayton
Carol, and Elwood

Preface

This book describes more than 1500 emulsifying agents which are currently available for industrial usage. The book will be of value to technical and managerial personnel involved in the specification and use of emulsifiers. It has been compiled from information received from numerous manufacturers and distributors of these products.

Emulsifying agents find uses in industries ranging from food preparation and processing to drilling fluids, from cosmetics and pharmaceuticals to heavy duty cleaners and degreasers for metal finishing. Textile manufacturers, pulp and paper processors, and agricultural product manufacturers also account for large quantities of emulsifiers. Another important application is emulsion polymerization systems which yield coatings, polishes, sealants and adhesives. Obviously, the market for these products is broad and new applications are constantly being developed.

Types of raw materials used for emulsifying agents include alkoxylates, alkanolamides, fatty acid esters, imidazolines, lecithins, quaternaries, phosphate esters, natural gums, glycerides, and sulfonates.

The data included represent selections from manufacturers' descriptions, in the manufacturer's own words, made at no cost to, nor influence from, the makers or distributors of the materials. Only the most recent information has been included. It is believed that all of the products listed here are currently available, which will be of utmost interest to readers concerned with product discontinuances.

The book lists the following product information, as available, in the manufacturer's own words:

(1) Company name and product category,

viii *Preface*

(2) Trade name and product number,

(3) Product description including properties and applications, as presented by the supplier.

Products are presented by company, and the companies are listed alphabetically. The table of contents is organized in such a way as to serve as a subject index to the book. Also included are a Chemical Name Index and a Trade Name Index, for easy and rapid location of products by the reader. In addition, another section, which will be useful, contains the Suppliers' Addresses. It can be found immediately following the Product Information section.

My fullest appreciation is expressed to the companies and organizations which supplied the data included in the book.

The reader may also be interested in another, possibly complementary, volume published by Noyes, entitled *Industrial Surfactants*.

January, 1990   Ernest W. Flick

NOTICE

To the best of our knowledge the information in this publication is accurate; however, the Publisher does not assume any responsibility for the accuracy or completeness of, or consequences arising from, such information. This industrial guide does not purport to contain detailed user instructions, and by its range and scope could not possibly do so. Mention of trade names or commercial products does not constitute endorsement or recommendation for use by the Publisher.

In some cases emulsifying agents could be toxic, and therefore due caution should be exercised. Final determination of the suitability of any information or product for use contemplated by any user, and the manner of that use, is the sole responsibility of the user. We strongly recommend that users seek and adhere to a manufacturer's or supplier's current instructions for handling each material they use.

The Author and Publisher have used their best efforts to include only the most recent data available. The reader is cautioned to consult the supplier in case of questions regarding current availability.

Contents and Subject Index

PRODUCT INFORMATION 1
  Akzo Chemicals Inc. 2
    ETHOMEEN Ethoxylated Aliphatic Amines 2
    ETHOQUAD Ethoxylated Quaternary Ammonium Salts 4
    ETHOFAT Ethoxylated Fatty Acids 4
    NEO-FAT Fatty Acids 5
    Nonionic Surfactants 6
      ARMID Aliphatic Amides 6
      ETHOMID Ethoxylated Amides 6
  Alcolac 7
    ABEX Surfactants 7
    CYCLOCHEMS 8
    DERMALCARES 10
  Alkaril Chemicals Ltd. 12
    ALKAMOX-Amine Oxides 12
    ALKAMULS AG–Agricultural Surfactants 13
    ALKAMULS–PEG Mono and PEG Di Fatty Acid Esters 15
    ALKAMULS–Sorbitan Esters 18
    ALKAMULS–Glycerol Esters 19
    ALKAMULS PS–Sorbitan Ester Ethoxylates 20
    ALKAPHOS–Phosphate Esters 22
    ALKASURF–Alcohol Ethoxylates 25
    ALKASURF–Fatty Acid Ethoxylates 30
    ALKASURF CO–Castor Oil Ethoxylates 32
    ALKASURF–Sulphonates 33
    ALKASURF LA-EP and BA-PE–Alcohol Alkoxylates 34
    ALKASURF NP–Nonyl Phenol Ethoxylates 36
    ALKASURF OP–Octyl Phenol Ethoxylates 39
    ALKAZINE–Imidazolines 41

Amerchol Corp. 42
    Emulsifiers 42
        GLUCAM E-20 Distearate 42
        GLUCATE DO 42
        OHLAN Hydroxylated Lanolin 43
        PROMULGEN 43
    SOLULANS 44
American Cyanamid Co. 48
    AEROSOL Surfactants 48
        Diester Sulfosuccinates 48
        Mono-Ester Sulfosuccinates 52
        Alkyl Naphthalene Sulfonate 52
        Nonyl Phenol Ether Sulfates 53
        Sulfosuccinates 54
        Dodecyl Diphenyl Oxide Sulfonate 54
        Naphthalene Formaldehyde Condensate 54
Angus Chemical Co. 55
    Amino Alcohols 55
        AMP 55
        TRIS AMINO 55
        AEPD 55
    Emulsifying Agents 56
        AMP-95 56
        DMAMP-80 56
BASF 57
    CREMOPHORS 57
Capital City Products Co. 59
    CAPMUL Mono- and Diglycerides 59
CasChem Inc. 60
    Emulsifiers–SURFACTOL Cationic Surfactants 60
    Emulsifiers–SURFACTOL Non-Ionic Surfactants 61
Central Soya 62
    Lecithins 62
        CENTROLEX 62
        CENTROLENE 62
        CENTROPHIL 62
        CENTROPHASE 62
        CENTROPHASE HR 62
        CENTROMIX 62
        CENTROL 62
        ACTIFLO 62
W.A. Cleary Products, Inc. 66
    CLEARATE Lecithin 66
Climax Performance Materials Corp. 67
    ACTRABASE Emulsifiers 67
    ACTRASOL Emulsifiers and Lubricants 67
Cyclo Chemicals Corp. 68
    CYCLOCHEM 68

Contents and Subject Index   xi

```
    Fatty Acid Esters ................................... 68
    Polyethylene Glycol Ester Group ...................... 69
    Blended Products .................................... 69
        CYCLOSHEEN 202 .................................. 70
    Other Available Products ............................. 70
DeSoto, Inc. ............................................. 71
    Esters and Ethoxylated Esters ........................ 71
        DESOTAN ......................................... 71
    Ethoxylated Amines ................................... 71
        DESOMEEN ........................................ 71
    Octyl Phenol Ethoxylates ............................. 72
        DESONIC ......................................... 72
    Linear Alcohol Ethoxylates ........................... 72
    Phosphate Esters ..................................... 73
        DESOPHOS ........................................ 73
    Castor Oil Ethoxylates ............................... 74
DuPont Co. ............................................... 75
    DUPONOL Alcohol Sulfates as Emulsifying Agents ....... 75
        AVITEX AD ....................................... 77
    Surfactants—Wetting Agents/Dispersing Agents/Emulsifying
    Agents ............................................... 78
        Anionic ......................................... 78
            Alcohol Sulfates, Sodium Salts .............. 78
            Alcohol Sulfates, Amine Salts ............... 78
            Alcohol Phosphates .......................... 79
                ZELEC ................................... 79
            Sulphonates, Aliphatic ...................... 79
                ALKANOL ................................. 79
                PETROWET ................................ 79
            Sulfonates, Alkylaryl ....................... 79
        Nonionic ........................................ 80
            Alcohol/Ethylene Oxide Adducts .............. 80
                MERPOL .................................. 80
        Specialties ..................................... 80
        Fluorosurfactants ............................... 81
            Anionic ..................................... 81
                ZONYL ................................... 81
            Nonionic .................................... 81
            Amphoteric .................................. 81
Eastman Chemical Products, Inc. .......................... 82
    Food-Grade Emulsifiers ............................... 82
        MYVEROL Distilled Monoglycerides ................ 82
        MYVACET Distilled Acetylated Monoglycerides ..... 83
        MYVAPLEX Concentrated Glyceryl Monostearates .... 84
        MYVATEM Dispersing Agents ....................... 84
        MYVEROL P-06 Distilled Propylene Glycol Monoester  84
        MYVEROL SMG Succinylated Monoglycerides ......... 84
    MYVATEX Food Emulsifier Blends ....................... 85
```

DO CONTROL Strengthener (K) 86
LIQUID LITE Food Emulsifier (K)......................86
MIGHTY SOFT Softener (K)............................86
MONOSET Food Emulsifier K..........................87
TEXTURE LITE Food Emulsifier (K)....................87
FMC Corp. 88
  Carrageenan Stabilizer/Emulsifier 88
Goldschmidt Chemical Corp. 89
  Emulsifiers for the Preparation of Acid-, Alkaline- and Salt-Stable
    Emulsions 89
    TEGINACID 89
    EMULGATOR 90
  Products for Pharmaceutical Preparations—Emulsifiers 91
    TEGIN 91
  Products for Pharmaceutical Preparations—Solubilizers/Co-Emulsifiers 92
    TAGAT. 92
Grindsted Products, Inc. 93
  Emulsifiers..93
    AMIDAN. 93
    ARTODAN 93
    CETODAN. 93
    DIMODAN. 93
    EMULDAN 93
    FAMODAN 94
    LACTODAN 94
    LIPODAN 94
    PANODAN 94
    PROMODAN 94
    TRIODAN..94
Gumix International 95
  Agar Agar 95
  Guar Gum 95
  Gum Arabic 96
  Gum Tragacanth.......................................97
  Locust Bean Gum 98
Harcros Chemicals Inc....................................99
  Emulsifiers..99
    T-MULZ 99
    CASUL..101
  T-MULZ Phosphate Esters...............................102
Henkel Corp. 104
  Emulsifiers...104
    CUTINA 104
    DEHYMULS 104
    EUMULGIN..104
    GENEROL...105
    LANETTE...105

Contents and Subject Index   xiii

Hercules Inc. 106
   HERCULES AR150 and AR160 Surfactants 106
Humko Chemical Division 107
   ATMOS and ATMUL Glycerol Esters 107
   Food Emulsifiers 108
      Mono- and Diglyceride Emulsifiers 108
      Synergistic Emulsifiers Blends 110
         TANDEM 110
   HYSTRENE Lauric and Myristic Acids 112
   INDUSTRENE Stearic and Palmitic Acids 113
   1,3-Propylenediamines 114
      KEMAMINE 114
   Sorbitan Esters 115
ICI Specialty Chemicals 116
   Emulsification Technology 116
   HYPERMER Polymeric Surfactants 118
Jordan Chemical Co. 122
   Phosphate Esters 122
      JORDAPHOS 122
Lanaetex Products, Inc. 125
   Emulsifiers 125
      ANATOL 125
      ETHLANA 125
      ETHOXOL 126
      ETHOXYCHOL 128
      EXTAN 128
      LANAETEX 130
      LANOBASE 132
      LANOLA 132
      LANOXIDE 133
      LANTOX 133
      LANYCOL 134
      LAXAN 135
      LINSOL 137
      NAETEX 137
Lipo Chemicals Inc. 138
   Emulsifiers 138
      LIPOSORB 138
      LIPOCOL 146
      LIPOPEG 156
      LIPO 163
      LIPOMULSE 165
      LIPOLAN 166
      LIPOWAX 166
      LIPAMIDE 169
Lonza Inc. 170
   LONZEST Sorbitan Esters 170
Mayco Oil & Chemical Co. 172

EMULAMID Lubricant-Emulsifiers... 172
MAYSOL Emulsifier Bases ... 173
Mazer Chemicals... 174
AVANEL S Surfactants ... 174
MACOL Nonionic Surfactants... 176
MACOL Fatty Alcohol Ethers... 176
MACOL and MAZAWAX Cosmetic Emulsifiers... 184
MAFO Amphoteric Surfactants ... 185
MAPEG Polyethylene Glycol Esters ... 187
MAZAMIDE Alkanolamides ... 194
MAZAWET Wetting Agents... 199
MAZEEN Ethoxylated Amines ... 200
MAZOL Glycerol and Polyglycerol Esters ... 202
MAZON Ethoxylated Sorbitol Esters ... 204
MAZOX Amine Oxides... 205
S-MAZ Sorbitan Fatty Acid Esters ... 206
T-MAZ Ethoxylated Sorbitan Fatty Acid Esters ... 209
Milliken Chemicals... 212
SYN FACS... 212
Mobay Chemical Corp.... 215
Emulsifier K30 ... 215
NL Chemicals... 216
KELECIN F and 1081 Surfactants ... 216
Patco Products ... 217
ALPHADIM 90AB Distilled Monoglyceride ... 217
ALPHADIM 90NLK Distilled Monoglyceride ... 218
ALPHADIM 90SBK... 219
DOUBLE SOFT Concentrated Hydrate ... 220
VERV... 220
EMPLEX Sodium Stearoyl Lactylate... 221
STARPLEX 90 Water Dispersible Distilled Monoglyceride... 222
Quantum Chemical Corp.... 223
EMEREST Glycerol Esters ... 223
EMID Alkanolamides... 225
EMSORB Ethoxylated Sorbitan Esters ... 226
EMSORB Sorbitan Fatty Acid Esters ... 228
Ethoxylated Fatty Acids and Polyethylene Glycol Fatty Acid Esters ... 230
TRYDET... 232
TRYCOL Ethoxylated Alcohols... 237
TRYCOL Ethoxylated Alkylphenols... 242
TRYFAC Phosphate Esters ... 247
TRYLOX Ethoxylated Sorbitol and Ethoxylated Sorbitol Esters ... 250
TRYLOX Ethoxylated Triglycerides... 251
TRYMEEN Ethoxylated Fatty Amines ... 253
Miscellaneous Surfactants ... 255
TRYLON... 255
EMERY... 255

Sandoz Chemicals 256
   CHEMICAL 39 BASE 256
   CHEMICAL BASE 6532 257
   SANDOPAN Carboxylated Surfactants 258
ScanRoad Inc. 260
   CATIMULS Emulsifiers 260
Shell Chemical Co. 262
   NEODOLS 262
   NEODOL Ethoxylates 263
Werner G. Smith, Inc. 266
   Non Ionic Emulsifier T-9 266
Sonneborn Division 268
   BARIUM PETRONATE Oil-Soluble Petroleum Sulfonate 267
   CALCIUM PETRONATE Oil-Soluble Petroleum Sulfonate 268
   SODIUM PETRONATE Oil-Soluble Petroleum Sulfonate 269
Stepan Co. 270
   Agricultural Products 270
      TOXIMUL 270
      STEPFAC 273
      NIPOL 275
      MICRO-STEP 275
   Esters 276
      Emulsifiers, Opacifiers 276
      Pearlescent Agents, Auxiliary Emulsifiers 277
      Emulsifiers, Viscosity Builders 278
   Personal Care Products—Cationics 279
      Stearyl Dimethyl Benzyl Ammonium Chloride 279
         AMMONYX 279
      Cetyl Trimethyl Ammonium Chloride 279
   Personal Care Products—Nonionics/Amphoterics 280
      Ethoxylated Alkanolamides 280
         AMIDOX 280
      Amine Oxides 280
Unichema Chemicals, Inc. 281
   ESTOL Esters 281
      Oleates 281
      Caprylates, caprates 282
      Stearates 283
      Laurates, Myristates, Palmitates 284
      Others 284
      Acetates 285
   PRICERINE Glycerines 286
Van den Bergh Food Ingredients Group 287
   DUR-EM Emulsifiers 287
   DUR-EM Mono- and Diglycerides—Kosher 289
   DURFAX Emulsifiers 290
   Emulsifiers 291
      DURLAC 291

                DUR-LO 292
                DURTAN 293
                DURKEE. 293
                DURPRO. 293
                ICE. 294
            SANTONE Emulsifiers. 295
            TALLY 100 Dough Conditioners and Softeners 296
        Westvaco Corp.. 297
            Asphalt Chemicals. 297
                INDULIN 297
                POLYFON. 297
            Rubber Chemicals 299
                WESTVACO. 299
        Witco Corp.. 300
            Surfactants for Emulsion Polymerization—General Emulsifier
            Recommendations. 300
                WITCOLATE 300
                WITCONATE 300
                EMCOL. 301
                EMPHOS. 301
                WITCONOL. 301

SUPPLIERS' ADDRESSES 303

CHEMICAL NAME INDEX. 307

TRADE NAME INDEX 321

PRODUCT INFORMATION

AKZO CHEMICALS INC.: ETHOMEEN Ethoxylated Aliphatic Amines:

ETHOMEEN polyethoxylated amines are cationic in character but, with progressive ethoxylation, they become increasingly non-ionic and water-soluble.

Ethoxylated amines are used as neutralizing agents for acrylate thickeners, as emulsifiers, and as dye leveling, wetting and anti-static agents.

Trade Name:

ETHOMEEN C/12:
    CTFA Adopted Name: PEG-2 Cocamine
    Form: Liquid
    Conc (%): 99

ETHOMEEN C/15:
    CTFA Adopted Name: PEG-5 Cocamine
    Form: Liquid
    Conc (%): 99

ETHOMEEN C/20:
    CTFA Adopted Name: PEG-10 Cocamine
    Form: Liquid
    Conc (%): 99

ETHOMEEN C/25:
    CTFA Adopted Name: PEG-15 Cocamine
    Form: Liquid
    Conc.: 99

ETHOMEEN 18/12:
    CTFA Adopted Name: PEG-2 Stearamine
    Form: Solid
    Conc (%): 99

ETHOMEEN 18/15:
    CTFA Adopted Name: PEG-5 Stearamine
    Form: Solid
    Conc (%): 99

ETHOMEEN 18/20:
    CTFA Adopted Name: PEG-10 Stearamine
    Form: Liquid to Paste
    Conc (%): 99

ETHOMEEN 18/25:
    CTFA Adopted Name: PEG-15 Stearamine
    Form: Liquid to Paste
    Conc (%): 99

AKZO CHEMICALS INC.: ETHOMEEN Ethoxylated Aliphatic Amines (Continued):

Trade Name:

ETHOMEEN 18/60:
   CTFA Adopted Name: PEG-50 Stearamine
   Form: Paste to Solid
   Conc (%): 99

ETHOMEEN O/12:
   CTFA Adopted Name: PEG-2 Oleamine
   Form: Liquid
   Conc (%): 99

ETHOMEEN O/15:
   CTFA Adopted Name: PEG-5 Oleamine
   Form: Liquid
   Conc (%): 99

ETHOMEEN O/25:
   CTFA Adopted Name: PEG-15 Oleamine
   Form: Liquid
   Conc (%): 99

ETHOMEEN S/12:
   CTFA Adopted Name: PEG-2 Soyamine
   Form: Viscous Liquid
   Conc (%): 99

ETHOMEEN S/15:
   CTFA Adopted Name: PEG-5 Soyamine
   Form: Liquid
   Conc (%): 99

ETHOMEEN S/20:
   CTFA Adopted Name: PEG-10 Soyamine
   Form: Liquid
   Conc (%): 99

ETHOMEEN S/25:
   CTFA Adopted Name: PEG-15 Soyamine
   Form: Liquid
   Conc (%): 99

ETHOMEEN T/12:
   CTFA Adopted Name: PEG-2 Tallow Amine
   Form: Paste
   Conc (%): 99

ETHOMEEN T/15:
   CTFA Adopted Name: PEG-5 Tallow Amine
   Form: Liquid to Paste
   Conc (%): 99

ETHOMEEN T/25:
   CTFA Adopted Name: PEG-15 Tallow Amine
   Form: Liquid to Paste
   Conc (%): 99

AKZO CHEMICALS INC.: ETHOQUAD Ethoxylated Quaternary Ammonium Salts:

These products are used as emulsifying conditioners and plasticizing agents in hair sprays.

Trade Name:

ETHOQUAD C/12:
   CTFA Adopted Name: PEG-2 Cocomonium Chloride
   Form: Liquid
   Conc (%): 73

ETHOQUAD C/25:
   CTFA Adopted Name: PEG-15 Cocomonium Chloride
   Form: Liquid
   Conc (%): Liquid

ETHOQUAD O/12H:
   CTFA Adopted Name: PEG-2 Oleamonium Chloride
   Form: Liquid
   Conc (%): 72

ETHOQUAD 18/12:
   CTFA Adopted Name: PEG-2 Stearmonium Chloride
   Form: Paste
   Conc (%): 70

ETHOQUAD 18/25:
   CTFA Adopted Name: PEG-15 Stearmonium Chloride
   Form: Liquid
   Conc (%): 95

ETHOFAT Ethoxylated Fatty Acids:

These ethoxylated fatty acids are recommended for use as emulsifiers and emollients for creams, lotions and cleansers.

Trade Name:

ETHOFAT C/25:
   CTFA Adopted Name: PEG-15 Cocoate
   Form: Paste
   Conc (%): 97

ETHOFAT O/20:
   CTFA Adopted Name: PEG-10 Oleate
   Form: Liquid
   Conc (%): 97

AKZO CHEMICALS INC.: NEO-FAT Fatty Acids:

These fatty acids are useful in the manufacture of shaving creams, hand creams, soaps and lotions. They act as emulsifiers and emollients and serve as intermediates for the production of esters of use in many cosmetic applications:

Trade Name:

NEO-FAT 8:
   CTFA Adopted Name: Caprylic Acid
   Form: Liquid
   Conc (%): 100

NEO-FAT 10:
   CTFA Adopted Name: Capric Acid
   Form: Solid
   Conc (%): 100

NEO-FAT 12-43:
   CTFA Adopted Name: Lauric Acid
   Form: Flake
   Conc (%): 100

NEO-FAT 255:
   CTFA Adopted Name: Coconut acid
   Form: Solid
   Conc (%): 100

NEO-FAT 265:
   CTFA Adopted Name: Coconut Acid
   Form: Paste
   Conc (%): 100

NEO-FAT 16:
   CTFA Adopted Name: Palmitic Acid
   Form: Flake
   Conc (%): 100

NEO-FAT 18:
   CTFA Adopted Name: Stearic Acid
   Form: Flake
   Conc (%): 100

NEO-FAT 18-55:
   CTFA Adopted Name: Stearic Acid
   Form: Flake
   Conc (%): 100

AKZO CHEMICALS INC.: Nonionic Surfactants:

ARMID Aliphatic Amides:
ETHOMID Ethoxylated Amides:

The ARMID series of aliphatic amines comprises inert waxes. They enhance the lubricity and mold release characteristics of lipsticks and improve emollient properties in creams and antiperspirants.

Ethoxylated amides can be used as emulsifiers in creams or lotions and as foam stabilizers and conditioning additives in shampoos.

Trade Name:

ARMID C:
    CTFA Adopted Name: Cocamide
    Form: Flake
    Conc (%): 99

ARMID HT:
    CTFA Adopted Name: Hydrogenated Tallow Amide
    Form: Flake
    Conc (%): 99

ARMID 18:
    CTFA Adopted Name: Stearamide
    Form: Flake
    Conc (%): 99

ARMID O:
    CTFA Adopted Name: Oleamide
    Form: Flake
    Conc (%): 99

ETHOMID HT/23:
    CTFA Adopted Name: PEG-13 Tallow Amide
    Form: Solid
    Conc (%): 99

ETHOMID HT/60:
    CTFA Adopted Name: PEG-50 Tallow Amide
    Form: Pellet or Paste
    Conc (%): 99

ETHOMID O/17:
    CTFA Adopted Name: PEG-7 Oleamide
    Form: Liquid
    Conc (%): 99

ALCOLAC: ABEX Surfactants:

ABEX 22-S:
Anionic Surfactant
ABEX 22-S is an outstanding surfactant for vinyl monomer emulsion polymerization, providing the combined advantages of anionic and nonionic systems in one molecule.

Typical Properties:
   Form: Pale yellow liquid
   Active Ingredient, percent: 25
   Viscosity @ 25C, cps: 1,000

ABEX 22-S is used in the manufacture of emulsion polymer latices for floor polish, latex paints, paper coatings, pigment binders, and adhesives.

ABEX 26-S:
Anionic Surfactant
ABEX 26-S is an anionic surfactant intended for emulsion polymerization applications where relatively large particle size is desirable.

Typical Properties:
   Form: Pale yellow liquid
   Active Ingredient, percent: 33
   Viscosity @ 25C, cps: 100

ABEX 26-S should be used in the emulsion polymerization of acrylics, vinyl acetate and other vinyl monomers.

ABEX 33S:
Anionic Surfactant

Typical Properties:
   Active, %: 27.0
   Viscosity (cps @ 25C): 300

Properties:
   High Foaming
   Good Wetting
   Good Dispersing
   Excellent lime-soap dispersant

Applications:
   Foaming/froth agent for textile resin finishes where compatability with salts and polymers is necessary.
   Base for scrub soaps, rug shampoos, general detergent applications.
   Emulsifier for vinyl and acrylic polymerization.

ALCOLAC: CYCLOCHEMS:

CYCLOCHEM NI:
Emollient Emulsifying Wax

Typical Properties:
  Physical Appearance: White to Off White Flakes
  Ionic Character: Nonionic
  Odor: Mild
  CTFA Identification: Cetearyl Alcohol (and) Ceteareth-20

CYCLOCHEM NI is a versatile, emulsifying wax custom developed for use in the preparation of exquisite cosmetic products.

CYCLOCHEM PEG 200 MO:
Polyethyelene Glycol 200 Monooleate

Appearance: Clear Oily Liquid
Saponification Value: 120-130
Odor: Mild, Fatty

CTFA Designation: PEG 4 Oleate

CYCLOCHEM PEG 200 MS:

Polyethylene Glycol 200 Monostearate

Appearance: Soft White Wax
Melting Point (C): 31 Typical
Saponification Value: 120-129
Odor: Mild, Characteristic

CYCLOCHEM PEG 200 MS is one of the more popular fatty acid esters in commercial use. It is commonly used in creams and lotions as an auxiliary emulsifier.
Additional esters in this family are produced from polyglycols in the 300 to 600 molecular weight range.

CTFA Designation: PEG-4 Stearate

CYCLOCHEM PEG 400 DL:

Polyethyelene Glycol 400 Dilaurate

Appearance: Clear Liquid
Saponification Value: 127-137
Odor: Mild, Characteristic

CTFA Designation: PEG-8 Dilaurate

ALCOLAC: CYCLOCHEMS(Continued):

CYCLOCHEM PEG 400 DS:
  Polyethylene Glycol 400 Distearate

Appearance: Soft White Wax
Saponification Value: 118-128
Odor: Mild, Characteristic

Commonly used in creams and lotions, as well as in creme rinses.
Additional esters in this family are produced from polyglycols in the 200, 300 and 600 molecular weight range.

CTFA Designation: PEG-8 Distearate

CYCLOCHEM PEG 600 ML:
  Polyethylene Glycol 600 Monolaurate

Appearance: Clear Liquid above 23C
Saponifification Value: 64-74
Odor: Mild, Characteristic

CTFA Designation: PEG-12 Laurate

CYCLOCHEM PEG 600 MO:
  Polyethylene Glycol 600 Monooleate

Appearance: Light Straw Colored Viscous Liquid
Saponification Value: 60-69
Odor: Mild Characteristic

CTFA Designation: PEG-12 Oleate

CYCLOCHEM POL:
  Nonionic Emulsifying Wax

Physical Appearance: Waxy White Solid
Ionic Character: Nonionic
Saponification Value: 11.0-16.0
CTFA Identification: Cetearyl Alcohol (and) Ceteth-20 (and)
                     Glycol Stearate

CYCLOCHEM POL is a nonionic, self-emulsifying wax ideally suited for the preparation of oil in water type creams and lotions.

ALCOLAC: DERMALCARES:

DERMALCARE EGMS/SE:
An anionic self-emulsifying wax that creates creams and lotions of the oil in water type with excellent feel on the skin.

Appearance: Off White Flakes
Melting Point: 55-60C
Saponification Value: 168-175
Odor: Mild, Characteristic

Commonly used for moisturizing lotions and creams as well as pharmaceutical preparations.

CTFA Designation: Glycol Stearate SE

DERMALCARE GMS/SE:
Glyceryl Monostearate - Self-Emulsifying

Appearance: Off White Flakes
Ionic Character: Nonionic/Anionic
Melting Point (C): 57-62
Saponification Value: 138-148

Based on glyceryl monostearate pure and a true fatty acid soap to give its self emulsifying character.

CTFA Designation: Glyceryl Stearate SE

DERMALCARE GTIS:
Isostearic Acid Triglyceride

Physical Appearance: Light Colored Liquid
Ionic Character: Nonionic
Saponification Value: 170-190
CTFA Identification: Triisostearin

Custom synthesized, cosmetic grade tryglyceride ideally suited for the preparation of unique personal care products.

DERMALCARE LVL:
Lauryl Lactate

Physical Appearance: Light Colored Liquid
Ionic Character: Nonionic
Saponification Value: 193-207
Melting Point (C): 4
CTFA Identification: Lauryl Lactate

Novel, cosmetic grade fatty ester specially synthesized to impart an exotic, satin-like feel to the skin.

ALCOLAC: DERMALCARE(Continued):

DERMALCARE MM/M:
  Myristyl Myristate

Physical Appearance: White to Off White Flake
Ionic Character: Nonionic
Odor: Mild
Saponification Value: 125-135
CTFA Identification: Myristyl Myristate

Cosmetic grade, fatty ester specially developed for use in the preparation of elegant skin care products.

DERMALCARE MST:

Appearance: White to Off White Flakes
Melting Point (C): 42-46
Saponification Value: 110-125
Molecular Structure: Typical for Spermaceti
Odor: Bland

Specially formulated wax ester similar to spermaceti in physical and chemical characteristics.

CTFA Name: Myristyl Stearate

DERMALCARE SDG:
  PEG-2 Stearate

Physical Form: White Solid Wax
Melting Range: 42-45C
Acid Value: 5.0 maximum
Iodine Value: 0.3 maximum

DERMALCARE SDG is a neutral grade diethylene glycol monostearate suitable for use as a thickener, opacifier and auxiliary emulsifier for shampoos and cosmetic creams of both W/O and O/W types.

ALKARIL CHEMICALS LTD.: ALKAMOX-Amine Oxides:

The ALKAMOX line of amine oxides is derived from high molecular weight tertiary amines through a hydrogen peroxide oxidation process.

The amine oxides are useful in shaving creams, lotions and fine fabric laundry formulations. The ALKAMOX Amine Oxides are of interest in other applications that include petroleum fuel additives, plating bath additives, antistats for textile finishing, emulsion polymerisation initiators and pigment dispersants.

These notably mild, high sudsing products provide foam stability comparable to such stabilizers as the alkanolamides in liquid detergent and liquid toiletrie preparations. In shampoo and bath products, the ALKAMOX Amine Oxides are mild to the skin, impart lubricity and emolliency, have antistatic and conditioning effects on the hair, are resistant to hard water precipitation and have good lime soap dispersing properties.

Products:

ALKAMOX ODM:
Substantive to skin and hair thereby providing antistat emolliency and conditioning features to shampoo formulations.
Appearance: clear liquid
% Amine Oxide: 50

ALKAMOX L20:
ALKAMOX LO:
Excellent foamers, wetters and foam stabilizers for rug shampoos, fine laundry detergents, dishwashing detergents, shampoos, bubble baths, cleaner formulations and antistatic textile softeners. Its antistatic properties together with its natural detergency and emolliency properties make it an alternative to alkanolamides in certain applications. It also finds acceptance in other areas; notably, as a foam stabilizer in foam rubber, in electroplating paper coatings and as a pour point depressant for mineral oils.
Appearance: clear liquid
% Amine Oxide: 29-31

ALKAMOX CAPO:
Exhibits stronger foam stabilization characteristics than ALKAMOX LO, consequently it is particularly suited to hair shampoo formulations. Applications are otherwise similar to ALKAMOX LO.
Appearance: clear liquid
% Amine Oxide: 29-31

ALKARIL CHEMICALS LTD.: ALKAMULS AG - Agricultural Surfactants:

The ALKAMULS AG Series is comprised of a wide range of agricultural emulsifiers and adjuvants, which are designed for use with herbicides, insecticides and fungicides marketed as emulsifiable concentrates, invert concentrates, wettable powders, flowables and aqueous concentrates.

Products:

ALKAMULS AG-100 and AG-200 Series:
  Emulsifier matched pairs for stable toxicants especially organo phosphate insecticides and non-saponifiable herbicides.
   Type: Calcium sulfonate-nonionic ethoxylate blends
   Appearance: liquids to flowable pastes

ALKAMULS AG-300 and AG-400 Series:
   Emulsifiers for phenoxy ester herbicides.
   Type: amine sulfonate-nonionic ethoxylate blends
   Appearance: liquids to flowable pastes

ALKAMULS AG-700 Series:
   Emulsifiers for easily degraded toxicants
   Type: nonionic ethoxylate blends
   Appearance: liquids

ALKASURF NP-6, 8, 9, 10, 12,
        OP-6, 8, 10, 12
        LA-7, 9, 11, 16
   Adjuvants for addition to toxicant concentrates.
   Type: nonionic ethoxylates
   Appearance: liquids to solids

ALKAMULS AG-821 and AG-826:
   Premium emulsifiers and adjuvants for 83/17 crop oil/surfactant concentrates.
   Type: ethoxylated ester nonionic blends
   Appearance: clear amber liquids

ALKAMULS AG-824 and AG-827:
   Emulsifiers and adjuvants for 83/17 crop oil/surfactant concentrates.
   Type: nonionic ethoxylates
   Appearance: liquids

ALKAMULS AG-825:
   Used as emulsifier and adjuvant for 98/2 crop oil/surfactant blends.
   Type: nonionic ethoxylate
   Appearance: clear liquid

ALKAMULS AG-823:
   Used as emulsifier and adjuvant for vegetable oil/surfactant concentrates.
   Type: nonionic ethoxylate
   Appearance: clear liquid

ALKARIL CHEMICALS LTD.: ALKAMULS AG - Agricultural Surfactants (Continued):

Products:

ALKAMULS AG-900, AG-902, AG-906,
    AG-913, AG-914, AG-915:
  Spreading and wetting agents for aqueous pesticide systems.
  Type: nonionic ethoxylate blends
  Appearance: clear liquids

ALKAMULS AG-903:
  Foaming adjuvant with wetting, spreading and sticking properties.
  Type: anionic-nonionic blend
  Appearance: clear liquid

ALKAMULS AG-901, AG-904, AG-918:
  Spreading and sticking agents
  Type: anionic-nonionic blends
  Appearance: liquids to pastes

ALKAMULS AG-923:
  Used as a soil wetting agent
  Type: nonionic ethoxylate
  Appearance: clear liquid

ALKARIL CHEMICALS LTD.: ALKAMULS - PEG Mono and PEG Di Fatty Acid Esters:

ALKAMULS Mono and Di Fatty Acid Esters are specialty esters of various fatty acids and polyethylene glycols. The numbers following the trade name designate the average molecular weight of the polyethylene glycol used. The letters following these numbers designate the fatty acid (hydrophobe) used.

These mono ester emulsifiers vary from oil soluble to water soluble and are used in many different applications that range from cutting oils, solvent cleaners, degreasers and pesticide formulations. They are used in the tanning and textile processes as lubricant-softeners.

The Di Fatty Acid Esters are generally more oil soluble and lower foaming than their mono ester counterparts. For these reasons they are generally used as emulsifiers for oils especially in cases where foam can create problems.

Products:

ALKAMULS 400-MO:
This moderately water soluble surfactant is used in the textile industry as a dyeing assistant and in the leather industry as an emulsifier for neats-foot oil fat liquors.
  HLB: 11.0
  Hydrophobe: oleic
  Appearance: yellow liquid
  Sap Value: 77-87

ALKAMULS 600-MO:
This water soluble surfactant is used as a co-emulsifier with ALKASURF 0-7 in various industrial applications.
  HLB: 13.0
  Hydrophobe: oleic
  Appearance: yellow liquid
  Sap Value: 57-67

ALKAMULS EGMS:
Opacifying and pearlescing agent for liquid cosmetic and detergent compounds.
  HLB: 2.9
  Hydrophobe: stearic
  Appearance: white solid
  Sap Value: 174-184

ALKARIL CHEMICALS LTD.: ALKAMULS - PEG Mono and PEG Di Fatty Acid Esters(Continued):

Products:

ALKAMULS 200-MS:
 Emulsifiers for fats and oils. It provides softening and lubricating properties to textiles and leather.
 HLB: 8.0
 Hydrophobe: stearic
 Appearance: white solid
 Sap Value: 120-130

ALKAMULS 400-MS:
 Functions as a self emulsifying lubricant and softener in textile compositions designed for synthetic fibres.
 HLB: 11.2
 Hydrophobe: stearic
 Appearance: white solid
 Sap Value: 80-90

ALKAMULS 200-ML:
 Emulsifier, coupling agent and solubilizer in metal working fluids, industrial lubricants and textile lubricants.
 HLB: 9.8
 Hydrophobe: lauric
 Appearance: yellow liquid
 Sap Value: 142-152

ALKAMULS 400-ML:
 Emulsifier and co-emulsifier for various cosmetic and toiletry preparations. Defoamer and leveling agent for latex paints. Dispersant for pigment and dye systems.
 HLB: 12.8
 Hydrophobe: lauric
 Appearance: yellow liquid to paste
 Sap Value: 91-101

ALKAMULS 400-DS:
 ALKAMULS 400-DS is an oil soluble wax-like emulsifier and thickener used in cosmetic and industrial emulsions.
 HLB: 7.8
 Hydrophobe: stearic
 Appearance: cream solid
 Sap Value: 115-130

ALKARIL CHEMICALS LTD.: ALKAMULS - PEG Mono and PEG Di Fatty Acid Esters(Continued):

Products:

ALKAMULS 600-DS:
The PEG-Distearates are generally used as lubricants and softeners in textile applications, and as opacifiers and emulsifiers in cosmetic preparations.
  HLB: 10.6
  Hydrophobe: stearic
  Appearance: cream solid
  Sap Value: 96-106

ALKAMULS 400-DO:
  HLB: 7.2
  Hydrophobe: oleic
  Appearance: amber liquid
  Sap Value: 105-115

ALKAMULS 600-DO:
  HLB: 10.0
  Hydrophobe: oleic
  Appearance: amber liquid
  Sap Value: 92-100

ALKAMULS 400-DL:
  HLB: 10.0
  Hydrophobe: lauric
  Appearance: yellow liquid to paste
  Sap Value: 132-142

ALKAMULS 600-DL:
  HLB: 11.5
  Hydrophobe: lauric
  Appearance: yellow liquid to paste
  Sap Value: 106-116

ALKAMULS 400-GDL:
  HLB: 10.5
  Hydrophobe: lauric
  Appearance: amber liquid
  Sap Value: 125-135

ALKAMULS 600-DT:
  HLB: 10.0
  Hydrophobe: tallow
  Appearance: amber liquid
  Sap Value: 90-105

These lipophilic members are used as emulsifiers and solubilizers for mineral oils, fats and solvents; as emulsifiers for kerosene and agricultural chemical sprays; as emulsifiers in metal working fluids, industrial lubricants and textile lubricants; and as viscosity control additives in amphoteric toiletry preparations.

ALKAMULS 6000-DS:
Hydrophilic emulsifier and thickener used in textile printing, pigment manufacturing and cosmetics
  HLB: 18.4
  Hydrophobe: stearic
  Sap Value: 14-20

ALKARIL CHEMICALS LTD.: ALKAMULS - Sorbitan Esters:

The ALKAMULS Sorbitan Esters are lipophilic emulsifiers and coupling agents.

The ALKAMULS Sorbitan Esters are frequently used alone or in combination with the corresponding ethoxylated sorbitan esters for the preparation of both W/O and O/W emulsions. These esters are used in cosmetics, household products, emulsifiable concentrates and industrial oils.

Products:

ALKAMULS SML:
Sorbitan monolaurate is a water dispersible emulsifier for oils and fats in cosmetic and industrial oil products. It is also used to retard starch crystallization in jellies. Used as a lubricant antistat process aid in PVC resin manufacture.
HLB: 8.6
Hydrophobe: mono-laurate
Appearance: amber liquid
Sap Value: 160-170
Hydroxyl Value: 320-350

ALKAMULS SMO:
Sorbitan mono oleate is a versatile oil soluble emulsifier and coupling agent for medicants and for petroleum oils, fats and waxes in the industrial, cosmetic and textile industries. It also functions as a textile and leather lubricant and softener. SMO improves pigment dispersions in lipsticks, eyeliners, mascara and other coloured cosmetics. Oil based ointments include SMO to impart a smooth feel to the skin and to reduce greasiness.
HLB: 4.3
Hydrophobe: mono-oleate
Appearance: amber liquid
Sap Value: 145-160
Hydroxyl Value: 193-210

ALKAMULS SMS:
Sorbitan monostearate is used to prepare silicone defoamer emulsions for various industrial applications; paraffin wax emulsions for processing paper coatings and industrial oil emulsions. SMS serves as a textile process lubricant and as an internal lubricant for PVC film.
HLB: 4.7
Hydrophobe: mono-stearate
Appearance: cream flake
Sap Value: 147-157
Hydroxyl Value: 235-260

ALKARIL CHEMICALS LTD.: ALKAMULS - Sorbitan Esters(Continued):

Products:

ALKAMULS STO:
  Sorbitan trioleate is used to compound textile and leather softener finishes.
  HLB: 1.8
  Hydrophobe: tri-oleate
  Appearance: amber liquid
  Sap Value: 170-190
  Hydroxyl Value: 55-70
ALKAMULS STS:
  Sorbitan tristearate is an extremely hydrophobic emulsifier and finds application as a fibre-to-metal lubricant for synthetic and cotton fibres.
  HLB: 2.1
  Hydrophobe: tri-stearate
  Appearance: cream flake
  Sap Value: 176-188
  Hydroxyl Value: 66-80

ALKAMULS - Glycerol Esters:
The ALKAMULS Glycerol Esters are mixtures of mono and diesters of lauric, oleic and stearic fatty acids.
Products:

ALKAMULS GMO-45LG:
  HLB: 3
  Hydrophobe: mono-oleate
  Ester Conc.: 40% min. Mono-Ester
  Free Glycerine: 1% max.
ALKAMULS GDO:
  HLB: 2.6
  Hydrophobe: di-oleate
  Free Glycerine: 2% max.
  Frequently used in mold release agents as a rust prevention additive for compounded oils and as lubricant component in synthetic fibre spin finishes. GMO is used as a lubricant-antistat aid in processing PVC film.
ALKAMULS GTO:
  Glycerol trioleate in an emulsified form produces an excellent lubricant for textiles, leather and metals. The sulfated form is a useful textile and leather softener.
  HLB: 0.8
  Hydrophobe: tri-oleate
  Ester Conc.: 85% min. Tri-Ester
ALKAMULS GMS-45:
  Glycerol monostearate is an emulsifier, common to hand creams, lotions and other cosmetics formulations. GMS-45 also serves as a textile lubricant-softener.
  HLB: 3
  Hydrophobe: mono-stearate
  Ester Conc.: 40% min. Mono-Ester
  Free Glycerine: 4-7%

ALKARIL CHEMICALS LTD.: ALKAMULS PS - Sorbitan Ester Ethoxylates:

The ALKAMULS Polyoxyethylene Sorbitan Esters are emulsifiers and coupling agents. They are frequently used in combination with the lipophilic ALKAMULS Sorbitan Esters from which they are derived. The numeral in the product name designates the molar quantity of ethylene oxide.

The ALKAMULS Polyoxyethylene Sorbitan Esters are O/W emulsifiers and function as co-emulsifiers for petroleum oils, fats and solvents; as emulsifiers in cosmetics, household products, industrial lubricants, fibre to metal textile lubricants and softeners for fibre and yarn; as solubilizers for oils and fragrances.

Products:

ALKAMULS PSML-4:
ALKAMULS PSML-4 is used in PVC emulsion polymerisation processing.
HLB: 13.3
Hydrophobe: mono-laurate
Sap Value: 100-115
Hydroxyl Value: 215-255

ALKAMULS PSML-20:
This multipurpose o/w emulsifier is used extensively to solubilize vitamin oils, essential oils, balsam and tar preparations in cosmetics and pharmaceuticals and to solubilize fragrances in cosmetics.
ALKAMULS PSML-20 is widely used as a viscosity modifier in non-irritating shampoos. Used extensively in textiles as a nylon spin finish; as a processing rinse aid in rayon finishing and as an emulsifier for dye carriers.
HLB: 16.7
Hydrophobe: mono-laurate
Appearance: yellow liquid
Sap Value: 40-50
Hydroxyl Value: 96-108

ALKAMULS PSMO-5:
In the textile industry, top quality fibre lubricants and softeners are prepared from ALKAMULS PSMO-5 emulsified oils. Quality cutting oils are produced with 10% PSMO-5 in the oil phase.
HLB: 10.0
Hydrophobe: mono-oleate
Appearance: yellow liquid to paste
Sap Value: 96-104
Hydroxyl Value: 134-150

ALKARIL CHEMICALS LTD.: ALKAMULS PS - Sorbitan Ester Ethoxylates (Continued):

Products:

ALKAMULS PSMO-20:
   Functions as an emulsifier for aliphatic alcohols in tobacco sucker control concentrates. It is a versatile O/W emulsifier and functions as a co-emulsifier with sorbitan esters for petroleum oils, fats, solvents and waxes. It performs similar functions to that of ALKAMULS PSML-20.
   HLB: 15.0
   Hydrophobe: mono-oleate
   Appearance: yellow liquid
   Sap Value: 45-55
   Hydroxyl Value: 65-80
ALKAMULS PSMS-4:
   In pharmaceuticals, the waxy PSMS-4 is used in suppositories. It is a useful fibre-to-metal lubricant for all fibres and yarns.
   HLB: 9.6
   Hydrophobe: mono-stearate
   Appearance: cream solid
   Sap Value: 98-113
   Hydroxyl Value: 170-200
ALKAMULS PSMS-20:
   Used as an O/W emulsifier for mineral oils, fats and waxes. This emulsifier is used in the preparation of paraffin wax emulsions for textiles and paper coatings. It is a useful fibre-to-metal textile lubricant.
   HLB: 15.0
   Hydrophobe: mono-stearate
   Appearance: yellow liquid to paste
   Sap Value: 45-55
   Hydroxyl Value: 81-96
ALKAMULS PSTO-20:
   O/W emulsifier for petroleum oils, fats and waxes. It is an excellent textile and leather lubricant. It is an effective emulsifier for oils and fats used in textile finishes and fibreglass lubricants.
   HLB: 11.0
   Hydrophobe: tri-oleate
   Appearance: yellow liquid
   Sap Value: 80-95
   Hydroxyl Value: 39-52
ALKAMULS PSTS-20:
   The pronounced lubricating and softening property of ALKAMULS PSTS-20 makes it useful in textile processing and in compounding textile finishes. It is used to emulsify petroleum oils and vegetable oils.
   HLB: 10.5
   Hydrophobe: tri-stearate
   Appearance: cream solid
   Sap Value: 88-98
   Hydroxyl Value: 44-60

ALKARIL CHEMICALS LTD.: ALKAPHOS - Phosphate Esters:

This class of surfactants combines numerous important properties such as efficient oil emulsification, good detergency for cotton and synthetics, effective wetting properties, excellent rinsability and dispersing properties.

They are stable to extremes of acidity and alkalinity; soluble in aromatic and chlorinated solvents; help sequester iron and other metal ions and have low toxicity and good rust inhibition properties.

Products:

ALKAPHOS L3-64A:
Oil soluble emulsifier and wetting agent. Extremely effective antistat-detergent in dry cleaning charge soaps and is an excellent non-corrosive lubricant for industrial coolant formulations. Useful textile lubricant.
 Hydrophobe: aliphatic
 Appearance: yellow liquid
 Form: acid
 % Active: 100

ALKAPHOS R6-33A:
 Hydrophobe: aromatic
 Appearance: pale yellow liquid
 Form: acid
 % Active: 100

ALKAPHOS R6-33S:
 Hydrophobe: aromatic
 Appearance: pale yellow liquid
 Form: sodium salt
 % Active: 88 min

Oil soluble emulsifiers and wetting agents. Effective antisoil redeposition agents. Lubricants for textile fibres.

ALKAPHOS B6-56A:
Water dispersible emulsifier. The wetting, antistatic and non-corrosive properties makes it applicable in textile operations such as scouring, kier boiling, peroxide bleaching and as an antistatic lubricant for yarns. It is also used in dry cleaning charge soap formulations; solvent degreasers and as a lubricant in industrial cooloant formulations.
 Hydrophobe: mixed base
 Appearance: yellow liquid
 Form: acid
 % Active: 100

ALKARIL CHEMICALS LTD.: ALKAPHOS - Phosphate Esters(Continued):

Products:

ALKAPHOS R6-36A:
  Hydrophobe: aromatic
  Appearance: pale yellow liquid
  Form: acid
  % Active: 100

ALKAPHOS L6-15S:
  Hydrophobe: aliphatic
  Appearance: yellow liquid
  Form: potassium salt
  % Active: 90

ALKAPHOS R9-07A:
  Hydrophobe: aromatic
  Appearance: amber liquid
  Form: acid
  % Active: 100
  Oil soluble, high foaming, emulsifiers and detergents with excellent wetting properties.

ALKAPHOS R9-47A:
  Versatile detergent for compounding heavy duty liquid detergents and dry cleaning detergents. A useful dedusting agent for alkaline powder cleaners. In textiles, its low rewetting property finds applications in the water repellent treatment of cottons.
  Hydrophobe: aromatic
  Appearance: yellow liquid
  Form: acid
  % Active: 100

ALKAPHOS R6-15A:
  Low foaming hydrotrope. Effective at solubilizing low foaming nonionic surfactants in liquid alkaline cleaners.
  Hydrophobe: aromatic
  Appearance: pale yellow liquid
  Form: acid
  % Active: 100

ALKAPHOS L4-27A:
  High foaming detergent particularly suited to solubilizing nonionic surfactants in cleaners containing moderate concentrations of caustic and electrolytes.
  Hydrophobe: aliphatic
  Appearance: pale yellow liquid
  Form: acid
  % Active: 100

ALKARIL CHEMICALS LTD.: ALKAPHOS - Phosphate Esters(Continued):

Products:

ALKAPHOS L3-15A:
Moderate foaming detergent for cleaners containing moderate concentrations of caustic and electrolytes.
    Hydrophobe: aliphatic
    Appearance: pale yellow liquid
    Form: acid
    % Active: 100

ALKAPHOS L6-36A:
This excellent emulsifier and wetting agent coupled with its powerful solubilization characteristics makes it particularly suited to compounding highly alkaline and high electrolyte content liquid cleaning preparations.
    Hydrophobe: aliphatic
    Appearance: yellow liquid
    Form: acid
    % Active: 80

ALKAPHOS L6-36S:
This neutral potassium salt of a hydrotropic phosphate ester is used to effectively solubilize low foam nonionic surfactants without contributing to additional foam height. Used in industrial cleaners to solubilize normal concentrations of alkaline salts.
    Hydrophobe: aliphatic
    Appearance: yellow liquid
    Form: potassium salt
    % Active: 41-43

ALKAPHOS R5-09A:
    Hydrophobe: aromatic
    Appearance: amber liquid
    Form: acid
    % Active: 100

ALKAPHOS R5-09S:
    Hydrophobe: aromatic
    Appearance: pale yellow liquid
    Form: potassium salt
    % Active: 50 min.
Combination of excellent coupling properties, low foaming and high caustic and electrolyte tolerances make these products particularly suited to highly alkaline, highly electrolytic low foaming cleaning preparations.

ALKARIL CHEMICALS LTD.: ALKASURF - Alcohol Ethoxylates:

This group of ALKASURF biodegradable surfactants represents a wide range of ethoxylates processed from various fatty alcohol hydrophobes. These extremely stable products are excellent detergents, wetting agents and emulsifiers; they are used as dispersants, solubilizers, coupling agents, fibre lubricants, antistats, levelling agents and dyeing assistants.

The lower POE ratio members are oil soluble detergents, emulsifiers and co-emulsifiers that will readily emulsify a wide range of oils, waxes and solvents. As intermediates, they can be sulfated to produce high foaming anionics for liquid detergent, shampoos and bubble baths.

The higher POE ratio members of this group function as water soluble emulsifiers that have many diverse applications including general purpose cleaners, heavy duty liquid and powder detergents, metal cleaners, mold release agents and as solvent emulsifiers in textile dye carrier systems.

The Oleyl and Stearyl Alcohol derivatives function mainly as emulsifiers and solubilizers in roll-on deodorants and topical cosmetic applications. The Tallow derivatives are used in the textile industry as detergents and lubricants and as dyeing assistants for wool/synthetic blends.

Products:

ALKASURF LA23-3:
  Detergent-emulsifier, intermediate, dispersant. A detergent intermediate for the manufacture of biodegradable ethoxysulfates for liquid hand washing, shampoo, bubble bath and specialty industrial preparations. Performs well as a wetter and co-emulsifier with ALKASURF LA23-6.5 in prespotting and prelaundry wash treatment formulations.
  HLB: 8.1
  Hydrophobe: C12-C13
  Appearance: liquid

ALKASURF LA23-6.5:
  A general purpose detergent with good wetting properties, designed for household detergent products.
  HLB: 12.0
  Hydrophobe: C12-C13
  Appearance: liquid
  Cloud Point: 42-48

ALKARIL CHEMICALS LTD.: ALKASURF - Alcohol Ethoxylates (Continued):

Products:

ALKASURF LAN-1:
   HLB: 3.7
   Hydrophobe: C12-C14
   Appearance: liquid
   Cloud Point C: 37-39

ALKASURF LAN-2:
   HLB: 5.8
   Hydrophobe: C12-C14
   Appearance: liquid
   Cloud Point C: 50-54

ALKASURF LAN-3:
   HLB: 8.0
   Hydrophobe: C12-C14
   Appearance: liquid
   Cloud Point C: 58-62
   Natural C12-C14 fatty alcohol containing 1 mole, 2 mole and 3 mole E.O. per mole of alcohol respectively. Detergent intermediates for the manufacture of biodegradable shampoo grade ethoxysulfates.

ALKASURF LAN-23:
   Water soluble natural C12-C14 fatty alcohol ethoxylate which functions as an emulsifier and solubilizer in cosmetics; as a post ad stabilizer for synthetic latices and as a solvent emulsifier for textile dye carriers.
   HLB: 17.0
   Hydrophobe: C12-C14
   Appearance: white solid
   Cloud Point C: 90.5-93.5

ALKASURF LA-3:
   A detergent intermediate for the manufacture of high foaming biodegradable ethoxysulfates for liquid dishwashing detergent, bubble baths, shampoos, and miscellaneous industrial applications. ALKASURF LA-3 is an excellent W/O emulsifier for mineral oil, kerosene and chlorinated solvents; these mineral oil emulsions find use as textile lubricants.
   HLB: 8.0
   Hydrophobe: C12-C15
   Appearance: liquid
   Cloud Point: 58-62

ALKARIL CHEMICALS LTD.: ALKASURF - Alcohol Ethoxylates
(Continued):

Products:

ALKASURF LA-7:
  Water soluble biodegradable surfactant used in a variety of household and institutional cleaner formulations. The strong wetting and emulsifying properties makes it versatile in both aqueous and kerosene based pre-spotter formulations.
  HLB: 12.0
  Hydrophobe: C12-C15
  Appearance: liquid
  Cloud Point C: 48-52

ALKASURF LA-9:
  Water soluble biodegradable detergent. General purpose nonionic emulsifier suited to many applications, but particularly to heavy duty controlled foaming detergents and heavy duty liquid detergents in which LA-9 retains its high performance over a wide range of temperature.
  HLB: 13.1
  Hydrophobe: C12-C15
  Appearance: white paste
  Cloud Point: 75-79

ALKASURF LA-9, 85%:
  Liquid, dilution of ALKASURF LA-9
  HLB: 13.1
  Hydrophobe: C12-C15
  Appearance: liquid
  Cloud Point: 75-79

ALKASURF LA-12:
  HLB: 14.4
  Hydrophobe: C12-C15
  Appearance: white paste
  Cloud Point C: 90-100

ALKASURF LA-12, 80%:
  HLB: 14.4
  Hydrophobe: C12-C15
  Appearance: liquid
  Cloud Point C: 90-100

ALKASURF LA-15:
  HLB: 15.3
  Hydrophobe: C12-C15
  Appearance: white solid
  Cloud Point C: 85-88
  Versatile nonionic emulsifiers similar to LA-9 but they are more hydrophilic. LA-12 has been shown to be effective in the deresination of unbleached sulfite pulp.
    LA-12, 80% is a liquid, dilution of ALKASURF LA-12.

ALKARIL CHEMICALS LTD.: ALKASURF - Alcohol Ethoxylates (Continued):

Products:

ALKASURF TDA-5:
This water dispersible surfactant has outstanding wetting, emulsifying, and dispersing properties suited to low temperature textile scouring. This intermediate can be phosphated to produce phosphate esters.
  HLB: 10.5
  Hydrophobe: C13
  Appearance: liquid
  Cloud Point C: 65-68

ALKASURF TDA-6:
  HLB: 11.4
  Hydrophobe: C13
  Appearance: opaque liquid
  Cloud Point C: 69-73

ALKASURF TDA-7:
  HLB: 12.1
  Hydrophobe: C13
  Appearance: liquid
  Cloud Point C: 72-75

ALKASURF TDA-8.5:
  HLB: 12.5
  Hydrophobe: C13
  Appearance: liquid
  Cloud Point C: 52-56
These extremely stable high quality emulsifiers are used in topical cosmetic applications similar to the ALKASURF OA-10 series. The low unsaturation of the stearyl alcohol ethoxylates offer long shelf life stability and function over a wide pH range.

ALKASURF DA-4:
  HLB: 10.5
  Hydrophobe: C10
  Appearance: liquid
  Cloud Point C: 60-66

ALKASURF DA-6:
  HLB: 12.5
  Hydrophobe: C10
  Appearance: liquid
  Cloud Point C: 60-70
Excellent wetting penetrant and low foam characteristics find application in textile industry products. ALKASURF DA-6 finds use as emulsifier in polyethylene emulsions.

ALKARIL CHEMICALS LTD.: ALKASURF - Alcohol Ethoxylates (Continued):

Products:

ALKASURF OA-10:
   Typically used as the high HLB component of an emulsifier pair in topical cosmetic formulations where high quality, blandness and low odour are prerequisites.
   HLB: 12.4
   Hydrophobe: oleyl
   Appearance: paste
   Cloud Point C: 53-60

ALKASURF SA-2:
   These extremely stable high quality emulsifiers are used in topical cosmetic applications similar to ALKASURF OA-10. The low unsaturation of the stearyl alcohol ethoxylates offer long shelf life stability and function over a wide pH range.
   HLB: 4.9
   Hydrophobe: stearyl
   Appearance: white solid
   Cloud Point: 57-61

ALKASURF SA-10:
   HLB: 12.4
   Hydrophobe: stearyl
   Appearance: white solid
   Cloud Point: 60-64

ALKASURF TA-40:
   This strongly hydrophilic tallow fatty alcohol derivative is economical and more frequently used in the textile industry as detergent and as dyeing assistant for wool/synthetic fibre blends.
   HLB: 17.4
   Hydrophobe: tallow
   Appearance: white solid
   Cloud Point C: 75-79

ALKARIL CHEMICALS LTD.: ALKASURF - Fatty Acid Ethoxylates:

The ALKASURF Fatty Acid Ethoxylates are nonionic surfactants having a wide range of detergent and emulsifying properties.

Products:

ALKASURF O75-7:
 HLB: 10.0
 Hydrophobe: oleic
 Appearance: amber liquid
 Sap Value: 95-105

ALKASURF O-9:
 HLB: 11.0
 Hydrophobe: oleic
 Appearance: yellow liquid
 Sap Value: 75-85

ALKASURF O75-9:
 HLB: 11.0
 Hydrophobe: oleic
 Appearance: amber liquid
 Sap Value: 75-85

ALKASURF P-7:
 HLB: 10.2
 Hydrophobe: palmitic
 Appearance: yellow paste
 Sap Value: 94-111

ALKASURF PEL-9:
 HLB: 12.6
 Hydrophobe: pelargonic
 Appearance: clear to hazy liquid
 Sap Value: 94-104
 These moderately water soluble emulsifiers are used primarily in textile products as dyeing assistants and in particular as textile fibre additives based on their outstanding lubricant and antistat properties. They also find use in the leather industry as an emulsifier in fat liquors.

ALKASURF O-14:
 This water soluble surfactant is used as a co-emulsifier with O-7 and O-9 various industrial applications.
 HLB: 13.0
 Hydrophobe: oleic
 Appearance: yellow liquid
 Sap Value: 57-67

ALKARIL CHEMICALS LTD.: ALKASURF - Fatty Acid Ethoxylates (Continued):

Products:

ALKASURF S-1:
   Opacifying and pearlescing agent for liquid cosmetic and detergent compounds.
   HLB: 2.9
   Hydrophobe: stearic
   Appearance: white solid
   Sap Value: 174-184

ALKASURF S-8:
   HLB: 11.2
   Hydrophobe: stearic
   Appearance: white solid
   Sap Value: 80-90

ALKASURF S-9:
   HLB: 11.2
   Hydrophobe: stearic
   Appearance: white solid
   Sap Value: 80-90
   Functions as a self emulsifying lubricant and softener in textile compositions designed for synthetic fibres.

ALKASURF S65-8:
   Similar to ALKASURF S-8 but suited to applications where slightly higher melting point is required.
   HLB: 11.2
   Hydrophobe: stearic
   Appearance: white flake
   Sap Value: 80-95

ALKASURF S-40:
   This strongly hydrophilic surfactant performs as a lubricant and as an emulsifier in the production of concentrated, pourable textile lubricants and softeners based on waxy fatty esters. Widely used as an emollient and co-emulsifier in cosmetic creams and lotions.
   HLB: 17.0
   Hydrophobe: stearic
   Appearance: white flake
   Sap Value: 25-35

ALKASURF L-9:
   HLB: 12.8
   Hydrophobe: lauric
   Appearance: yellow liquid to paste
   Sap Value: 91-101

ALKASURF L-14:
   HLB: 14.6
   Hydrophobe: lauric
   Appearance: yellow liquid to paste
   Sap Value: 67-77
   Emulsifiers and co-emulsifiers for various cosmetic and toiletry preparations. Defoamers and leveling agents for latex paints. Dispersants for pigment and dye systems.

ALKARIL CHEMICALS LTD.: ALKASURF CO-Castor Oil Ethoxylates:

The ALKASURF CO series represent a range of ethoxylated castor oils. These ethoxylated triglycerides vary from lipophilic to hydrophilic in character and are designated by a number that approximates the average number of moles of E.O. per mole of castor oil.

The lipophilic members are soluble in chlorinated and aromatic solvents and are used as emulsifiers for waxes and oils.

The higher ethoxylates are frequently used as water soluble emulsifiers for oils, solvents and waxes. They are also used as lubricants in fat liquor baths found in the leather industry.

The most important applications are found in the textile industry where they are used as lubricants and antistats in textile fibre processing, emulsifiers for dye carrier and emulsifiers for hydrophobic glycerides in blended fibre lubricants.

Products:

ALKASURF CO-10:
In latex paints, it is used as pigment dispersant and simultaneously improves gloss and cold temperature stability. In textiles, it is used as emulsifier in low foam dye carriers.
HLB: 7.2
Cloud Point C: 45-49
SAP Value: 121-126

ALKASURF CO-15:
In textiles, this water dispersible emulsifier is used in synthetic fibre lubricants and as a vat dyeing assistant. Also used as a component of cutting oils and hydraulic fluids.
HLB: 8.6
Cloud Point: 56-60
SAP Value: 104-109

ALKASURF CO-20:
HLB: 10.3
Cloud Point: 65-69
SAP Value: 91-96

ALKASURF CO-25:
HLB: 11.0
Cloud Point: 46-50
SAP Value: 81-86

ALKASURF CO-30:
HLB: 11.7
Cloud Point: 48-61
SAP Value: 73-78

Water soluble emulsifiers for animal and vegetable fats and oils, fatty acids, waxes and solvents. They are excellent pigment dispersants. In leather processing, they are used for degreasing and as emulsifiers/lubricants in fat liquor baths. In textiles, they are used as emulsifiers in softener and dye carrier compositions and as emulsifiers/antistats in synthetic fibre lubricants

ALKARIL CHEMICALS LTD.: ALKASURF CO-Castor Oil Ethoxylates (Continued):

Products:

ALKASURF CO-36:
   HLB: 12.7
   Cloud Point C: 56-67
   SAP Value: 63-73

ALKASURF CO-40:
   HLB: 13.0
   Cloud Point C: 60-70
   SAP Value: 55-65
Water soluble emulsifiers, emollients and lubricants. Slightly more hydrophilic than CO-30 but properties and applications are essentially similar.

ALKASURF-Sulphonates:

The ALKASURF Sulphonates are derivatives of ALKASURF LA Acid, a linear alkyl benzene sulphonic acid.

ALKASURF IPAM:
The isopropylamine derivative is an oil soluble emulsifier frequently used to formulate solvent degreasers, emulsion cleaners and dry cleaning charge soaps. Added to fuel oil, it solubilizes trace water thereby eliminating corrosive water deposits.
   Description: isopropylamine DBS
   % Active: 90-93

ALKASURF T:
The triethanolamine derivative is a high sudsing, completely water soluble intermediate suited to bubble bath and concentrated shampoo compositions.
   Description: triethanolamine DBS
   % Active: 56-58

ALKASURF CA:
The calcium derivative is completely biodegradable and is used extensively as an emulsifier in combination with ethoxylated nonionics in self-dispersing liquids. In the textile industry it is used as a dispersant in the polyester yarn dyeing process. Emulsifier for agricultural chemicals.
   Description: calcium DBS
   % Active: 59-61

ALKARIL CHEMICALS LTD.: ALKASURF LA-EP and BA-PE - Alcohol Alkoxylates:

A series of biodegradable nonionic surfactants based on a linear alcohol hydrophobe and incorporating a modified polyoxyethylene hydrophile. These products offer one advantage of being more fluid when compared to the corresponding linear fatty alcohol ethoxylates. Generally, they tend not to gel when added to water.

Products:

ALKASURF LA-EP15:
A low foaming biodegradable surfactant recommended for use in automatic dishwasher formulations, metal cleaners, and household and for industrial formulations where low foam coupled with good detergency and rinse properties are desirable.
    HLB: 11.9
    Appearance: liquid
    % Active Min.: 99

ALKASURF LA-EP16:
Detergent wetting agent, emulsifier and dispersant used in various industrial and household cleaning products. A unique liquid product i.e. non gelling in aqueous dilutions, instantly soluble in cold water in all proportions.
    HLB: 13.0
    Appearance: liquid
    % Active Min.: 99

ALKASURF LA-EP25:
    HLB: 7
    Appearance: liquid
    % Active Min.: 99
ALKASURF LA-EP25LF:
    HLB: 7
    Appearance: liquid
    % Active Min.: 99
ALKASURF LA-EP35:
    HLB: 9
    Appearance: liquid
    % Active Min.: 99
ALKASURF LA-EP38:
    HLB: 9
    Appearance: liquid
    % Active Min.: 99
ALKASURF LA-EP45:
    HLB: 10
    Appearance: liquid
    % Active Min.: 99

Water soluble liquid biodegradable nonionic surfactants offering low foaming characteristics combined with excellent wetting. These products find application in industrial and household cleaning products especially machine dishwashing compounds and dishwasher rinse aids.

ALKARIL CHEMICALS LTD.: ALKASURF LA-EP and BA-PE - Alcohol Alkoxylates(Continued):

Products:

ALKASURF LA-EP59:
A liquid biodegradable nonionic surfactant which is readily soluble in cold water in all proportions. Applicable to formulating all purpose cleaners, heavy duty liquid and granular cleaners and detergent sanitizers. LA-EP59 is especially useful in car wash formulations both liquid and powder.
  HLB: 10.0
  Appearance: liquid
  % Active Min.: 99

ALKASURF LA-EP65:
A water soluble liquid biodegradable nonionic surfactant which is readily soluble in cold water in all proportions. Applications similar to LA-EP59 but offering increased water solubility at higher use temperatures.
  HLB: 11.5
  Appearance: liquid
  % Active Min: 99

ALKASURF LA-EP73:
A water soluble liquid biodegradable nonionic surfactant with applications similar to ALKASURF LA-EP59 but offering increased water solubility at higher use temperatures and salt levels.
  HLB: 14.0
  Appearance: liquid
  % Active Min.: 99

ALKASURF BA-PE70:
  HLB: 16.1
  Appearance: soft solid
  % Active Min.: 99

ALKASURF BA-PE80:
  HLB: 26.1
  Appearance: white solid
  % Active Min.: 99
Moderate foaming, water soluble emulsifiers for chlorinated solvents and as a component in emulsifier systems for herbicide and insecticide concentrates.

BA-PE80 is a particularly useful surfactant in applications at elevated temperatures. Used in the manufacture of particular iodophors and germicidal cleaners.

ALKARIL CHEMICALS LTD.: ALKASURF NP - Nonyl Phenol Ethoxylates:

The ALKASURF NP Ethoxylates have physical and chemical properties similar to the ALKASURF Ethoxylated Alcohols and ALKASURF Ethoxylated Octyl Phenols. The numeral in the product name designates the approximate molar quantity of ethylene oxide. In addition to being oil soluble emulsifiers the low POE ratio members are intermediates which can be sulfated to produce high foaming anionic surfactants.

Products:

ALKASURF NP-1:
Oil soluble emulsifier and dispersing agent for petroleum oils. Co-emulsifier and retardant in hair color preparations. Used as a defoamer in combination with water soluble ethoxylates.
  HLB: 4.6
  Appearance: liquid
  % Active Min.: 99

ALKASURF NP-4:
  HLB: 9.0
  Appearance: liquid
  % Active Min: 99
ALKASURF NP-5:
  HLB: 10.0
  Appearance: liquid
  % Active Min.: 99
ALKASURF NP-6:
  HLB: 11.0
  Appearance: liquid
  % Active Min: 99
Oil soluble detergents and emulsifiers. Emulsifiers for a wide range of fats, oils and waxes. Intermediates for sulfation to produce high foaming anionic detergents and for phosphorylation to produce lubricants and antistatic agents.

ALKASURF NP-8:
  HLB: 12.0
  Appearance: liquid
  % Active Min.: 99
ALKASURF NP-9:
  HLB: 13.4
  Appearance: liquid
  % Active Min.: 99
Water soluble, versatile detergents, wetting agents and emulsifiers. In the textile industry for the processing of wool, cotton and synthetics, they are used in all phases of production that include scouring, warp sizing, carbonizing, and bleaching. Applicable to formulating many household and industrial cleaning compounds and antimicrobials such as detergent sanitizers. As emulsifiers, they cover a wide range of medium polarity oils and solvents and find use in agricultutral chemical preparations.

ALKARIL CHEMICALS LTD.: ALKASURF NP - Nonyl Phenol Ethoxylates (Continued):

Products:

ALKASURF NP-10:
  HLB: 13.5
  Appearance: liquid
  % Active Min.: 99

ALKASURF NP-11:
  HLB: 13.8
  Appearance: liquid
  % Active Min.: 99

ALKASURF NP-12:
  HLB: 13.9
  Appearance: liquid
  % Active Min.: 99
  Water soluble surfactants compatible with iodophors, quaternaries and phenolics. They function as detergents, detergent additives, solubilizers and dispersants. Can be used in similar applications to ALKASURF NP-9 while being more soluble at elevated temperatures.

ALKASURF NP-15:
  HLB: 15.0
  Appearance: liquid to paste
  % Active Min.: 99

ALKASURF NP-15 80%:
  HLB: 15.0
  Appearance: liquid
  % Active Min.: 79-81
  The increased solubility at elevated temperatures makes these surfactants with good wetting and penetrating properties effective in highly electrolyte formulations such as bottle washing compounds, metal cleaners and heavy duty alkaline cleaners.

ALKASURF NP-20:
  High temperature textile scouring agent. Effective in high concentrations of electrolytes. Co-emulsifier for oils, fats, waxes and solvents. Acts as de-emulsifier for petroleum oil emulsions.
  HLB: 16.0
  Appearance: white solid
  % Active Min.: 99

ALKARIL CHEMICALS LTD.: ALKASURF NP - Nonyl Phenol Ethoxylates (Continued):

Products:

ALKASURF NP-30:
    HLB: 17.1
    Appearance: white solid
    % Active Min.: 99

ALKASURF NP-30 70%:
    HLB: 17.1
    Appearance: liquid
    % Active Min.: 69-71
    Recommended as solubilizers and co-emulsifiers for highly polar substrates. Applications are similar to ALKASURF NP-20.

ALKASURF NP-35 70%:
    HLB: 17.4
    Appearance: liquid
    % Active Min.: 69-71

ALKASURF NP-40:
    HLB: 17.7
    Appearance: white solid
    % Active Min.: 99

ALKASURF NP-40 70%:
    HLB: 17.7
    Appearance: liquid
    % Active Min.: 69-71

ALKASURF NP-50 70%:
    HLB: 18.0
    Appearance: liquid
    % Active Min.: 69-71

ALKASURF NP-100:
    HLB: 18.0
    Appearance: white solid
    % Active Min.: 99
    Highly water soluble emulsifiers and stabilizers used in applications where maximum water solubility is required at normal and elevated temperatures. Used widely in emulsion polymerisation of synthetic latices for latex paint, floor finishes, paper coatings and textile finishes.

ALKARIL CHEMICALS LTD.: ALKASURF OP - Octyl Phenol Ethoxylates:

The ALKASURF OP series have physical and chemical properties similar to both the ALKASURF Ethoxylated Alcohols and the ALKASURF Ethoxylated Nonyl Phenols. The numeral in the product name designates the approximate molar quantity of ethylene oxide.

Products:

ALKASURF OP-1:
  Oil soluble emulsifier and dispersing agent for petroleum oils. Co-emulsifier and retardant in hair color preparations.
  HLB: 3.6
  Appearance: liquid
  % Active min.: 99

ALKASURF OP-5:
  HLB: 10.4
  Appearance: liquid
  % Active min.: 99

ALKASURF OP-6:
  HLB: 11.4
  Appearance: liquid
  % Active min.: 99
  Useful emulsifiers to improve the detergency, dispersibility and wetting action of non polar hydrocarbon solvents and oils. Find use in solvent emulsion cleaners and dry cleaning charge soaps. Effectively used as agricultural and floor finish emulsifiers.

ALKASURF OP-8:
  Water dispersible surfactant with applications similar to OP-10 but producing a lower foam level. Used in metal cleaners, acid cleaners, floor cleaners, and controlled foam powdered laundry detergents. Effective emulsifier for pesticide formulations such as wettable powders and emulsifiable concentrates.
  HLB: 12.5
  Appearance: liquid
  % Active min: 99

ALKASURF OP-10:
  High performance, water soluble surfactant used to improve the detergency and wetting properties of household and industrial cleaning formulations. This excellent hard surface detergent, stable to strong acid and alkalies makes it versatile for laundry compounds, metal cleaners, acid cleaners, detergent sanitizers, floor cleaner and liquid hand dishwashing detergents.
  HLB: 13.5
  Appearance: liquid
  % Active min.: 99

ALKARIL CHEMICALS LTD.: ALKASURF OP - Octyl Phenol Ethoxylates (Continued):

Products:

ALKASURF OP-12:
Being slightly more water soluble than OP-10, ALKASURF OP-12 is an effective multipurpose surfactant. It is particularly suited to alkyl benzene sulphonate, liquid detergents, quaternary sanitizers and all purpose cleaners where it functions as a detergent, stabilizer-coupler-hydrotrope and foam stabilizer for the system. Also used as a wetting agent for caustic soda and mineral acid cleaners.
  HLB: 14.5
  Appearance: liquid
  % Active min.: 99

ALKASURF OP-30:
  HLB: 17.5
  Appearance: white solid
  % Active: 99

ALKASURF OP-30-70%:
  HLB: 17.5
  Appearance: liquid
  % Active min.: 69-71

ALKASURF OP-40:
  HLB: 18.0
  Appearance: white solid
  % Active min.: 99

ALKASURF OP-40-70%:
  HLB: 18.7
  Appearance: liquid
  % Active min.: 69-71

ALKASURF OP-70-50%:
  HLB: 18.7
  Appearance: liquid
  % Active min.: 49-51
Primary emulsifiers for vinyl acetate and acrylate emulsion polymerisation. These highly water soluble surfactants offer higher foaming characteristics than the lower ethoxylates in the series, making them useful in specialty applications.

ALKARIL CHEMICALS LTD.: ALKAZINE - Imidazolines:

ALKAZINE Fatty Imidazolines are specialty cationic emulsifiers. The letter following the ALKARIL trade name designates the hydrophobe.

In general, the ALKAZINES function as oil soluble emulsifying agents producing cationic O/W emulsions.

Products:

ALKAZINE C:
ALKAZINE C increases the lubricity of water soluble cutting oils and synthetic coolants. ALKAZINE C is an effective antistat and is used to treat woolen rugs, synthetic rugs, and plastic substrates to eliminate static charge accumulation. In the petroleum industry, ALKAZINE C is used in oil well acidifying and secondary recovery where its anticorrosion property and fungicidal action are important. It can also be used as an antifungal agent in the treatment of wood and is suggested as a slime control additive in paperboard.
   Hydrophobe: coconut
   Appearance: brown paste
   Imidazoline Content: 85% min.
   Neutralisation Equivalent: 265-295

ALKAZINE O:
ALKAZINE O is the most popular member of the series and functions as an emulsifier for both carnauba wax and light mineral oil in car wax emulsions. It is an effective wetting agent, emulsifier and an extremely effective corrosion inhibitor in toilet bowl cleaners prepared from hydrochloric acid. It is used for the flocculation of negatively charged particles in pigment flushing in the manufacture of paints and finds applications in emulsion cleaners and in agricultural emulsions. Also used to increase the lubricity of synthetic coolant formulations.
   Hydrophobe: oleic
   Appearance: amber liquid
   Imidazoline Content: 85% min.
   Neutralisation Equivalent: 335-365

ALKAZINE TO:
ALKAZINE TO is used as an emulsifier in the production of cationic bitumen emulsions. ALKAZINE TO can be used as an economical replacement for ALKAZINE O in many applications.
   Hydrophobe: tall oil
   Appearance: amber liquid
   Imidazoline Content: 80% min.
   Neutralisation Equivalent: 355-385

ALKAZINE TO-A:
Aggregate wetting agent and emulsifier in cationic asphalt emulsions.
   Hydrophobe: tall oil
   Appearance: amber liquid
   Imidazoline Content: 70% min.
   Neutralisation Equivalent: 240-300

AMERCHOL CORP.: Emulsifiers:

GLUCAM E-20 Distearate:

Esterified Ethoxylated Glucose Derivative Naturally-Derived Emollient and O/W Emulsifier
CTFA Adopted Name: Methyl Gluceth-20 Distearate--is a 100% active semisolid.
GLUCAM E-20 Distearate is an effective emollient and auxiliary oil-in-water emulsifier. Although many o/w nonionic emulsifiers are considered to be defatting, GLUCAM E-20 Distearate exhibits o/w emulsifier activity and behaves as a lubricant and emollient. This unusual combination of functional properties is attributed to the unique molecular structure of GLUCAM E-20 Distearate, particularly its derivation from methyl glucoside.
GLUCAM E-20 Distearate functions as an auxiliary o/w emulsifier by virtue of its molecular balance of the hydrophilic ethoxylated methyl glucoside and lipophilic distearate moieties.

Typical Analysis:
    Appearance: Yellow semisolid
    Acid Value: 3 max.
    Saponification Value: 66-80
    Hydroxyl Value: 58-75
    Heavy Metals: 20 ppm max.
    Arsenic: 2 ppm max.
    HLB (calculated): 12.5

GLUCATE DO:

Nonionic w/o Emulsifier Multifunctional Glucose Derivative
CTFA Adopted Name: Methyl Glucose Dioleate--is a 100% active, clear, viscous liquid with strong activity as a primary water-in-oil emulsifier. This unique product offers the cosmetic formulator a variety of benefits:
    - Efficient Performance
    - Multifunctionality
    - Safety
    - Glucose Chemistry

Typical Analysis:
    Appearance: Amber, viscous liquid
    Acid Value: 8 max.
    Saponification Value: 145-160
    Hydroxyl Value: 145-170
    Iodine Value: 60-75
    Arsenic: 2 ppm max.
    Heavy Metals: 20 ppm max.
    HLB (calculated): 5
    Flash Point (COC): 550F

AMERCHOL CORP.: Emulsifiers(Continued):

OHLAN Hydroxylated Lanolin:

W/O emulsifier, emulsion stabilizer and emollient for cosmetics and pharmaceuticals

OHLAN performs as a superior nonionic water-in-oil emulsifier, a stabilizer for oil-in-water systems, an emollient, and a pigment wetting and dispersing agent. It has an HLB value of 4.

CTFA Adopted Name: Hydroxylated Lanolin
Physical Description: Yellow-amber to light-tan waxy solid with mild characteristic odor.

Specifications:
  Acid Value: 10 max.
  Saponification Value: 95-110
  Hydroxyl value: 38-48
  Hydroxyl value-Stetzler: 60-85
  Iodine value (Hanus): 15-23
  Moisture: 0.25% max.
  Ash: 0.25% max.
  Melting Point: 39-46C
  Arsenic: 2 ppm max.
  Heavy Metals: 20 ppm max.

PROMULGEN:

Primary nonionic emulsifying and conditioning agents

PROMULGEN D:
  CTFA adopted name: Cetearyl Alcohol (and) Cetearath-20
  Chemical Description: Cetearyl alcohol and ethoxylated cetearyl alcohol
  Specifications:
    Melting point: 47 to 55C
    Saponification value: 2 max.
    Acid value: 1 max.
    Iodine value (Hanus): 2 max.

PROMULGEN G:
  CTFA adopted name: Stearyl Alcohol (and) Ceteareth-20
  Chemical Description: Stearyl alcohol and ethoxylated cetearyl alcohol

  Specifications:
    Melting Point: 55 to 63C
    Saponification value: 2 max.
    Acid Value: 1 max.
    Iodine value (Hanus): 2 max.

AMERCHOL CORP.: SOLULANS:

SOLULAN 5:
   Chemical Description: Ethoxylated (5 mole) complex of fatty and lanolin alcohols.
   Properties: Primary Water/Oil Emulsifier
      Auxiliary emulsifier and stabilizer for oil/water systems. Gives light lubricating feel. Shear thinning, improves gloss.
   Uses: W/O Creams & Lotions
         - Skin conditioner
         - Hair dressings
         - Cold creams
         - Cleansing creams
   HLB: 8
   Physical form and description: Amber semisolid
   CTFA Name: Laneth-5 (and) Ceteth-5 (and) Oleth-5 (and)
              Steareth-5

SOLULAN 16:
   Chemical Description: Ethoxylated (16 mole) complex of
                         fatty and lanolin alcohols
   Properties: Additive for soap/detergent systems
      Foam stabilizer versatile oil/water emulsifier and solubilizer. Provides soft, nontacky afterfeel.
   Uses: Conditioner, Liquid Soaps, Shampoos
         - Emollient for creams and lotions
         - Mousses
         - Solubilzer for PABA derivatives in shampoos
         - Shaving creams
         - Bar soaps
   HLB: 15
   Physical Form and Description: Light tan, waxy solid
   CTFA Name: Laneth-16 (and) Ceteth-16 (and) Oleth-16 (and)
              Steareth-16

SOLULAN 25:
   Chemical Description: Ethoxylated (25 mole) complex of
                         fatty and lanolin alcohols
   Properties: Primary oil/water emulsifier
      More hydrophilic than SOLULAN 16. Foam stabilizer. Soft, slightly waxy afterfeel.
   Uses: Creams & Lotions:
         - Foam stabilizer in detergent systems
         - Mousse products
         - Cleansing creams
   HLB: 16
   Physical Form and Description: Yellow waxy solid
   CTFA Name: Laneth-25 (and) Ceteth-25 (and) Oleth-25 (and)
              Steareth-25

AMERCHOL CORP.: SOLULANS(Continued):

SOLULAN C-24:
   Chemical Description: Ethoxylated (24 moles) cholesterol
                         fatty alcohol complex
   Properties: Very effective in preventing excessive viscosity
               buildup in fluid emulsions.
       Auxiliary oil/water emulsifier and stabilizer.
       Water-soluble cholesterol.
       Unusual waxy afterfeel.
   Uses: Anionic emulsions
         - Hair products for conditioning and cholesterol
           claims
   HLB: 14
   Physical Form and Description: Off-white to pale yellow
                                  waxy solid.
   CTFA Name: Choleth-24 (and) Ceteth-24

SOLULAN L-575:
   Chemical Description: Ethoxylated (75 moles) lanolin
   Properties: Conditioner for soap and detergent systems
       No effect on foaming characteristics. Leaves a soft,
nontacky afterfeel. Reduces defatting on the skin.
   Uses: Baby Wipes
         - Moisturizing/conditioning shampoos
         - Anti-cracking agent in soap bars
         - Skin cleansers
         - Creams & lotions
   HLB: 15
   Physical Form and Description: Amber fluid 50% solids in
water. (Also available as 100% waxy solid SOLULAN 75)
   CFTA Name: PEG-75 Lanolin

SOLULAN 97:
   Chemical Description: Acetylated complex of Polysorbate 80
                         and cetyl and lanolin alcohols
   Properties: Plasticizer for Polymers/Resins
       Ideal additive for non-aqueous systems. Nontacky,
lubricating afterfeel
   Uses: Hair Spray
         - Hair mousse products
         - Styling products
         - Hair sheen sprays
   HLB: 15
   Physical Form and Description: Light amber, viscous liquid
   CTFA Name: Polysorbate 80 Acetate (and) Cetyl Acetate (and)
              Acetylated Lanolin Alcohol

46 Emulsifying Agents

AMERCHOL CORP.: SOLULANS(Continued):

SOLULAN 98:
  Chemical Description: Partially acetylated complex of Polysorbate 80 and cetyl and lanolin alcohols
  Properties: Conditioner in aqueous detergent systems
    Primary oil/water emulsifier and solubilizer. Accelerates pearling of stearic acid emulsions. Versatile pigment dispersant in aqueous systems.
  Uses: Conditioning shampoo
      - Hair conditioner/rinse products
      - Wetting agent for benzoyl peroxide
      - Liquid makeup systems
      - Creams & lotions
  HLB: 13
  Physical Form and Description: Light amber, viscous liquid
  CTFA Name: Polysorbate 80 (and) Cetyl Acetate (and) Acetylated Lanolin Alcohol

SOLULAN PB-2:
  Chemical Description: Propoxylated (5 mole) lanolin alcohols
  Properties: Very effective pigment wetter & tackifier
    Compatible with ionic emulsions. Adds water resistance and emollience. Protective afterfeel. Adds gloss.
  Uses: Lipsticks
      - Pot lip gloss
      - Anhydrous liquid makeups
      - Mascaras
  HLB: 8 (Required)
  Physical Form: Amber, semisolid
  CTFA Name: PPG-2 Lanolin Alcohol Ether

SOLULAN PB-5:
  Chemical Description: Propoxylated (5 mole) lanolin alcohols
  Properties: Effective hair conditioner
    No adverse effect on foaming properties. Provides gloss and manageability.
  Hair Dressings: Anhydrous hair products such as:
          - Sheen sprays
          - Pomades
  HLB: 10 (Required)
  Physical Form and Description: Light amber, clear viscous liquid.
  CTFA Name: PPG-5 Lanolin Ether

AMERCHOL CORP.: SOLULANS(Continued):

SOLULAN PB-10:
  Chemical Description: Propoxylated (10 mole) lanolin alcohols
  Properties: Provides high film gloss
    Excellent resin plasticizer. Retains fluidity at low temperatures.
  Uses: Styling Products
    - Mousses
    - Hair sprays
    - Gels
    - W/O emulsions
  HLB: 12 (Required)
  Physical Form and Description: Straw, clear, liquid
  CTFA Name: PPG-10 Lanolin Alcohol Ether

SOLULAN PB-20:
  Chemical Description: Propoxylated (20 mole) lanolin alcohols
  Properties: Excellent spreading agent
    Excellent solvent for PABA esters in mineral oil. Soft and supple afterfeel. Pigment dispersant.
  Uses: Floating bath oils
    - Sunscreen emulsions
    - Sunscreen oils
    - Baby oils
    - Lip products
  HLB: 14 (Required)
  Physical Form and Description: Light straw, clear to hazy liquid
  CTFA Name: PPG-20 Lanolin Alcohol Ether

AMERICAN CYANAMID CO.: AEROSOL Surfactants:

Diester Sulfosuccinates:

AEROSOL Surfactants:

TR-70:
    Chemical Name: Sodium bistridecyl sulfosuccinate
    Type & Form: Anionic
                  70% Liquid in water & alcohol
    Biodegradability: Complete
    Surface Tension in Water (minimum) dynes/cm: 27
    FDA Approvals 21 CFR: 178.3400
    Features/Benefits: High oil solubility with limited water solubility. Surfactant for emulsion polymerization. Dispersant for pigments, dyes and polymers in hydrocarbon and other organic systems.

OT-75:
    Chemical Name: Sodium dioctyl sulfosuccinate
    Type & Form: Anionic
                  75% Liquid in water & alcohol
    Biodegradability: Complete
    Surface Tension in Water (minimum) dynes/cm: 26
    FDA Approvals 21 CFR: 178.3400
    Features/Benefits: Superior wetting and rewetting agent for increasing absorbency and penetration. Emulsifying agent, dewatering aid, surface tension reducer and surface property modifier. Surfactant for emulsion polymerization.

GPG:
    Chemical Name: Sodium dioctyl sulfosuccinate
    Type & Form: Anionic
                  70% Liquid in water & alcohol
    Biodegradability: Complete
    Surface Tension in Water (minimum) dynes/cm: 26
    FDA Approvals 21 CFR: 178.3400
    Features/Benefits: General purpose grade of AEROSOL OT for dust control, general wetting, emulsification and dispersion.

OT-70-PG:
    Chemical Name: Sodium dioctyl sulfosuccinate
    Type & Form: Anionic
                  70% Liquid in propylene glycol & water
    Biodegradability: Complete
    Surface Tension in Water (minimum) dynes/cm: 26
    FDA Approvals 21 CFR: 178.3400
    Features/Benefits: AEROSOL OT in propylene glycol and water. Designed as a very high flash point wetting agent and emulsifier. Features/Benefits--same as AEROSOL OT-75.

AMERICAN CYANAMID CO.: AEROSOL Surfactants(Continued):

Diester sulfosuccinates(continued):

OT-100:
   Chemical Name: Sodium dioctyl sulfosuccinate
   Type & Form: Anionic
                100% Waxy Solid
   Biodegradability: Complete
   Surface Tension in Water (minimum) dynes/cm: 26
   FDA Approvals 21 CFR: 178.3400
   Features/Benefits: Waxy solid form of AEROSOL OT. Used for
emulsification, dispersion, lubricating, wetting and water
displacement. Excellent mold release properties. Surface active
agent for water-free systems.

OT-B:
   Chemical Name: Sodium dioctyl sulfosuccinate
   Type & Form: Anionic
                85% active Powder
                15% sodium benzoate
   Biodegradability: Complete
   Surface Tension in Water (minimum) dynes/cm: 26
   FDA Approvals 21 CFR: 178.3400
   Features/Benefits: Adjuvant for agricultural chemical wettable
powders. Instantaneously soluble in water. Dispersing, wetting,
solubilizing agent, for improved color value, penetrant, surface
tension reducer. Pigment dispersing agent in plastics.

OT-S:
   Chemical Name: Sodium dioctyl sulfosuccinate
   Type & Form: Anionic
                70% Liquid in light petroleum distillate
   Biodegradability: Complete
   Surface Tension in Water (minimum) dynes/cm: 26
   FDA Approvals 21 CFR: 178.3400
   Features/Benefits: Solution of AEROSOL OT in high purity
light petroleum distillate. Excellent wetting and lubricating
agent. Superior detergent for oily stains. For use in organic
solvent systems.

OT-MSO:
   Chemical Name: Sodium dioctyl sulfosuccinate
   Type & Form: Anionic
                62% Liquid in mineral seal oil
   Biodegradability: Complete
   Surface Tension in Water (minimum) dynes/cm: 26
   FDA Approvals 21 CFR: 178.3400
   Features/Benefits: Same use as OT-S when a higher flash
point is required.

AMERICAN CYANAMID CO.: AEROSOL Surfactants(Continued):

Diester Sulfosuccinates(Continued):

MA-80:
   Chemical Name: Sodium dihexyl sulfosuccinate
   Type & Form: Anionic
               80% Liquid in water & alcohol
   Biodegradability: Slowly
   Surface Tension in Water (minimum) dynes/cm: 28
   FDA Approvals 21 CFR: 178.3400
   Features/Benefits: Penetrating agent, emulsifier, dispersing and solubilizing agent. Superior emulsifier for styrene-butadiene emulsion polymerization. High salt tolerance.

A-196:
   Chemical Name: Sodium dicyclohexyl sulfosuccinate
   Type & Form: Anionic
               85% Pellets in water
   Biodegradabilty: Complete
   Surface Tension in Water (minimum) dynes/cm: 39
   FDA Approvals 21 CFR: 178.3400
   Features/Benefits: For use in modified styrene butadiene emulsion polymerization. Imparts high surface tension, high CMC, promotes adhesion and reduces film water sensitivity properties.

A-196-40:
   Chemical Name: Sodium dicyclohexyl sulfosuccinate
   Type & Form: Anionic
               40% in water
   Biodegradability: Complete
   Surface Tension in Water (minimum) dynes/cm: 39
   FDA Approvals 21 CFR: 178.3400
   Features/Benefits: Solution form of AEROSOL A-196. Features/Benefits--same as AEROSOL A-196

AY-65:
   Chemical Name: Sodium diamyl sulfosuccinate
   Type & Form: Anionic
               65% Liquid in water & alcohol
   Biodegradability: Complete
   Surface Tension in Water (minimum) dynes/cm: 30
   FDA Approvals 21 CFR: 178.3400
   Features/Benefits: Wetting and dispersing agent. Water soluble. High electrolyte compatibility. Surfactant for emulsion polymerization.

AMERICAN CYANAMID CO.: AEROSOL Surfactants(Continued):

Diester Sulfosuccinates(Continued):

AY-100:
   Chemical Name: Sodium diamyl sulfosuccinate
   Type & Form: Anionic
             100% Waxy Solid
   Biodegradability: Complete
   Surface Tension in Water (minimum) dynes/cm: 30
   FDA Approvals 21 CFR: 178.3400
   Features/Benefits: Solid version of AEROSOL AY. Features/Benefits--same as AEROSOL AY-65. Used in water-free systems.

IB-45:
   Chemical Name: Sodium diisobutyl sulfosuccinate
   Type & Form: Anionic
             45% Liquid in water
   Biodegradability: Complete
   Surface Tension in Water (minimum) dynes/cm: 42
   FDA Approvals 21 CFR: 178.3400
   Features/Benefits: Surfactant/stabilizer in the emulsion polymerization of styrene-butadiene. Extremely hydrophilic, excellent electrolyte compatibility and dispersing properties.

Alkylamine-Guanidine Ethoxylate:

C-61:
   Chemical Name: Alkylamine-guanidine polyoxyethanol
   Type & Form: Cationic
             70% Liquid Paste
   Biodegradability: Partially
   Surface Tension in Water (minimum) dynes/cm: 40
   Features/Benefits: Pigment dispersant, flushing agent, wetting agent in plastics, paper, textiles and adhesive applications.

AMERICAN CYANAMID CO.: AEROSOL Surfactants(Continued):

Mono-Ester Sulfosuccinates:

A-102:
Chemical Name: Disodium ethoxylated alcohol half ester of sulfosuccinic acid
Type & Form: Anionic
        31% Liquid in water
Surface Tension in Water (minimum) dynes/cm: 33
FDA Approvals 21 CFR: 175.105
Features/Benefits: Solubilizer, foaming agent, dispersant, emulsifier amd surface tension reducer. High electrolyte tolerance. Generates small particle size vinyl acetate and acrylic polymer emulsions.

A-103:
Chemical Name: Disodium ethoxylated nonyl phenol half ester of sulfosuccinic acid.
Type & Form: Anionic
        34% Liquid in water
Surface Tension in Water (minimum) dynes/cm: 34
FDA Approvals 21 CFR: 175.105
Features/Benefits: Emulsifier, dispersant, foamer, solubilizer, limesoap dispersant, surface tension reducer. Compatible with divalent and trivalent cations.

A-268:
Chemical Name: Disodium isodecyl sulfosuccinate
Type & Form: Anionic
        50% Liquid in water
Surface Tension in Water (minimum) dynes/cm: 28
FDA Approvals 21 CFR: Petition Submitted
Features/Benefits: Emulsifier for emulsion and suspension polymerization of vinyl chloride, vinylidene chloride and co-monomers. Reduces yellowing of polymer when exposed to heat.

501:
Chemical Name: Proprietary composite (U.S. Patent 3,947,400)
Type & Form: Anionic
        50% Liquid in Water
Surface Tension in Water (minimum) dynes/cm: 28
FDA Approvals 21 CFR: Petition Submitted
Features/Benefits: Emulsifier for generation of small particle size vinyl acetate and acrylic polymer emulsions.

Alkyl Naphthalene Sulfonate:

OS:
Chemical Name: Sodium diisopropyl naphthalene sulfonate
Type & Form: Anionic ---- 75% active powder
Surface Tension in Water (minimum) dynes/cm: 35
FDA Approvals 21 CFR: 175.105
Features/Benefits: Stable wetting and dispersing agent.

AMERICAN CYANAMID CO.: AEROSOL Surfactants(Continued):

Nonyl Phenol Ether Sulfates:

NPES 458:
   Chemical Name: Ammonium salt of sulfated nonylphenoxy poly(ethyleneoxy)ethanol.
   Type & Form: Anionic
              58% Liquid in alcohol & water
   Surface Tension in Water (minimum) dynes/cm: 31
   FDA Approvals 21 CFR: 178.3400
   Features/Benefits: Surfactant for the emulsion polymerization of acrylic, styrene and vinyl acetate systems. High foaming surfactant for use in detergent, germicidal and textile applications.

NPES 930:
   Chemical Name: Ammonium salt of sulfated nonylphenoxy poly (ethyleneoxy) ethanol
   Type & Form: Anionic
              30% Liquid in water
   Surface Tension in Water (minimum) dynes/cm: 33
   FDA Approvals 21 CFR: 175.105/176.180
   Features/Benefits: Emulsifier for the generation of very fine particle size vinyl acetate, acrylic and styrene-acrylic latexes. Resulting films have superior water-resistant properties.

NPES 2030:
   Chemical Name: Ammonium salt of sulfated nonylphenoxy poly (ethyleneoxy) ethanol.
   Type & Form: Anionic
              30% Liquid in water
   Surface Tension in Water (minimum) dynes/cm: 43
   Features/Benefits: Same as AEROSOL NPES 930.

NPES 3030:
   Chemical Name: Ammonium salt of sulfated nonylphenoxy poly (ethyleneoxy) ethanol
   Type & Form: Anionic
              30% Liquid in water
   Surface Tension in Water (minimum) dynes/cm: 43
   FDA Approvals 21 CFR: 175.105/176.180
   Features/Benefits: Highly hydrophilic version of AEROSOL NPES 2030. Primary surfactant for emulsion polymerization of acrylic, vinyl acetate and styrene-acrylic systems. Very fine particle size emulsions. Forms films with superior water-resistance.

AMERICAN CYANAMID CO.: AEROSOL Surfactants(Continued):

Sulfosuccinates:

18:
    Chemical Name: Disodium N-octadecyl sulfosuccinamate
    Type & Form: Anionic ---- 35% Paste in water
    Biodegradabiliy: Complete
    Surface Tension in Water (minimum) dynes/cm: 39
    FDA Approvals 21 CFR: 176.170 & 176.180 with limitations
    Features/Benefits: Foaming agent for latex. Emulsifier, dispersant, detergent.

22:
    Chemical Name: Tetrasodium N-(1,2-dicarboxy-ethyl)-N-
              octadecyl sulfosuccinamate
    Type & Form: Anionic ---- 35% Liquid in water and alcohol
    Biodegradability: Complete
    Surface Tension in Water (minimum) dynes/cm: 41
    FDA Approvals: 178.3400 & 176.170 with limitations/175.105
    Features/Benefits: Emulsifier, dispersant, and hydrotrope/ solubilizer. Excellent salt tolerance and solubilizing power. Surfactant for emulsion polymerization of vinyl chloride and modified styrene-butadiene systems.

Dodecyl Diphenyl Oxide Sulfonate:

DPOS-45:
    Chemical Name: Disodium mono- & didodecyl diphenyl oxide disulfonate
    Type & Form: Anionic ---- 45% Liquid in water
    Surface Tension in Water (minimum) dynes/cm: 34
    FDA Approvals 21 CFR: 178.3400
    Features/Benefits: Emulsifying, dispersing and solubilizing agent exhibiting high electrolyte tolerance. Primary surfactant for emulsion polymerization systems. Resulting latexes have excellent mechanical and thermal stability. Extremely effective coupling agent.

Naphthalene Formaldehyde Condensate:

NS:
    Chemical Name: Sodium neutralized condensed naphthalene sulfuric acid.
    Type & Form: Anionic ---- 87% active powder
    Biodegradability: Slowly
    Surface Tension in Water (minimum) dynes/cm: 72
    FDA Approvals 21 CFR: 175.105/176.170/176.180
    Features/Benefits: Highly effective dispersant for pigments, extenders and fillers in aqueous media over a broad pH range.

ANGUS CHEMICAL CO.: Amino Alcohols:

AMP:
  2-Amino-2-methyl-1-propanol

TRIS AMINO:
  Tris(hydroxymethyl)aminomethane

AEPD:
  2-Amino-2-ethyl-1,3-propanediol

Product Specifications:

2-Amino-2-methyl-1-propanol
  AMP Regular:
  Neutral equivalent: 88.5-91
  Water, % by wt. (max.): 0.8

  AMP-95:
  Neutral equivalent: 93-97
  Water, % by wt. (max.): 5.8

2-Amino-2-ethyl-1,3-propanediol
  AEPD:
  Neutral equivalent: 121.5
  Water, % by wt. (max.): 3.8

Tris(hydroxymethyl)aminomethane
  TRIS AMINO:
  Crystals:
  Neutral equivalent: 121-122
  Water, % by wt. (max.): 0.5

  TRIS AMINO:
  40% Concentrate:
  Amine assay by titration, calc. as TRIS AMINO: 40+-2

Uses: In Emulsions:
   Amine soaps of fatty acids are widely used emulsifying agents. They appear in personal care products such as cosmetics and creams, home maintenance products such as floor polishes and cleaners, and industrial products such as insecticide sprays and asphalt emulsions. AMP-95 and DMAMP-80 form particularly efficient emulsifying agents which are far superior to those formed with the more traditional amines. The high base strength, low neutral equivalent, and low volatility of AMP make it possible to use less amine in emulsion formulations and still obtain better clarity, lower odor, and more stability than can be obtained with weaker or more fugitive amines.

ANGUS CHEMICAL CO.: Emulsifying Agents:

AMP-95:

AMP-95 (2-amino-2-methyl-1-propanol containing 5% water), when used in combination with fatty acids or nonionic surfactants, is a highly-effective emulsifying agent for emulsifiable waxes in aqueous systems. Emulsions of these waxes are used to impart outstanding physical properties to finished formulations of such diverse products as floor polishes, buffing compounds, drawing compounds, and textile lubricants.

When used in emulsions, AMP-95 provides a number of benefits:

Superior emulsion stability.
Light-colored emulsions.
Increased emulsion transparency.
Ready emulsification of a wide variety of waxes.
Excellent gloss and leveling of films.
Minimal fire hazard.
Ease of handling.

In addition, AMP-95 can be used to advantage in specific situations because of its high boiling point and excellent emulsification efficiency.

DMAMP-80:
80% 2-Dimethylamino-2-methyl-1-propanol solution

2-Dimethylamino-2-methyl-1-propanol, is the tertiary-amine homolog of 2-amino-2-methyl-1-propanol(AMP). 2-Dimethylamino-2-methyl-1-propanol is commercially available as DMAMP-80 which contains 20% by weight water.

Typical Properties:
  Neutral equivalent: ~148
  Specific Gravity at 25/25C: 0.95
  Weight per gallon at 25C: 7.9 lb
  Flash point, Tag open cup: 150F
              Tag closed cup: 153F
  Freezing point: -20C
  Boiling point at 760 mmHg: ~98C
  Viscosity at 25C, Gardner: A-A2
  pH of 0.1N aqueous solution: 11.6

Emulsifying Agent:
  DMAMP-80 in combination with an unsaturated fatty acid (e.g., oleic acid or tall oil fatty acids) is extremely effective in producing emulsions of waxes.

Corrosion Inhibition
Resin Solubilization
Catalyst
Additive

BASF: CREMOPHOR A Grades:

Nonionic emulsifying agents for the production of cosmetic and pharmaceutical oil-in-water emulsions.

Nature:
Nonionic emulsifying agents produced by reacting higher saturated fatty alcohols with ethylene oxide. CREMOPHOR A6 also contains free fatty alcohol.

Properties: The CREMOPHOR A grades are white, waxy or powdery substances. The powder is supplied in the form of free-flowing, non-dusting microbeads.

CREMOPHOR A6:

Appearance: White wax
Degree of ethoxylation: 6
Hydrophilic-lipophilic balance: 10-12
Saponification number: <3
Hydroxyl number: 115-135
Iodine number: <1
Acid number: <1
Dropping Point: 41-45C

CREMOPHOR A11:

Appearance: White wax
Degree of ethoxylation: 11
Hydrophilic-lipophilic balance: 12-14
Saponification number: <1
Hydroxyl number: 70-80
Iodine number: <1
Acid number: <1
Dropping point: 34-38C

CREMOPHOR A25:

Appearance: White microbeads
Degree of ethoxylation: 25
Hydrophilic-lipophilic balance: 15-17
Saponification number: <3
Hydroxyl number: 35-45
Iodine number: <1
Acid number: <1
Dropping point: 44-48C

CTFA name: Ceteareth-6
          Ceteareth-11
          Ceteareth-25

BASF: CREMOPHORS:

CREMOPHOR S9:
Polyethylene glycol 400 stearate for emulsifying oil-in-water preparations and stabilizing and thickening suspensions.

Nature:
Nonionic emulsifying agent produced by reacting stearic acid (1 mole) with ethylene oxide (9 moles).

Description:
CREMOPHOR S9 conforms to the monograph "Polyethylene glycol 400 stearate", German Pharmacopoeia, DAB 8.
CREMOPHOR S9 is a yellowish white, readily water-dispersible unctuous substance with a faint odour.
The HLB of CREMOPHOR S9 is approximately 12.

Specifications:
Saponification value: 88-98
Hydroxyl value: 80-105
Acid value: <2
Dropping point: 26-31C
Appearance: lighter GG 6
Ash: <0.2%
Foreign fatty acids: conforms

CREMOPHOR WO 7:
Water-in-oil emulsifier for cosmetic preparations. Especially suitable for the manufacture of soft and liquid emulsions.

Nature: Hydrogenated castor oil that has been made to react with 7 moles of ethylene oxide

CTFA Name: PEG-7 Hydrogenated Castor Oil

Synonyms: Polyethylene Glycol (7) Hydrogenated Castor Oil
Polyoxyethylene (7) Hydrogenated Castor Oil

Properties:
Appearance: Cloudy, yellowish, viscous liquid
Solubility 10%: Insoluble in water
Dispersible in liquid paraffin
Odour: Practically odourless

Specifications:
Saponification number: 125-150
Hydroxyl number: 100-130
Iodine colour number: <2
Acid number: <1
Water content (Fischer's method): <0.3%

Applications: Used in the manufacture of water-in-oil emulsions.

CAPITAL CITY PRODUCTS CO.: CAPMUL Mono- and Diglycerides:

CAPMUL mono- and diglycerides are nonionic emulsifiers made by reacting glycerine with fats, oils or fatty acids. They are lipophilic, insoluble in water and soluble in oils at elevated temperatures. They are used in fats and oils, often with other emulsifiers, to produce water in oil emulsions and to increase viscosity. All CAPMUL mono- and diglycerides use the highest quality raw materials in order to meet the rigid specifications of the cosmetic and pharmaceutical industries.

Product:

CAPMUL GDL:
  Form: Semi-Solid
  Iodine Value Max.: 20
  Sap. Value: 215-230
  M.P. C: 28-31
  HLB: 3.6

CAPMUL GMS:
  Form: Flake
  Iodine Value Max.: 5
  Sap. Value: 155-165
  M.P. C: 57-62
  HLB: 3.6

CAPMUL GMVS:
  Form: Plastic
  Iodine Value Max.: 75
  Sap. Value: 155-165
  M.P. C: 44-54
  HLB: 3.5

CAPMUL GMO:
  Form: Semi-Solid
  Iodine Value Max.: 75
  Sap. Value: 160-170
  M.P. C: 25 Max.
  HLB: 3.4

CAPMUL MCM:

CAPMUL MCM is a mono- and diglyceride of medium chain fatty acids (caprylic and capric). It has solvent properties and is useful as an oil-in-water emulsifier.

  Form: Liquid
  Acid Value Max.: 2.5
  Alpha Mono % Min.: 70

CASCHEM INC.: Emulsifiers--SURFACTOL Cationic Surfactants:

The SURFACTOL Q Series represents CASCHEM'S new and exciting line of Castor base quaterniums. These dimethyl amino propyl-amine quaterniums are prepared from high purity ricinoleic acid and hydroxystearic acid. They offer superior substantivity to skin and hair without build up. SURFACTOL Q series quaterniums offer the following advantages:

* Refatting agent.
* Impart substantivity and emolliency to skin and hair.
* Foam boosting properties in shampoos.
* Water soluble.
* Broad compatibility with anionic surfactants.
* Mildness.

Cationics:

SURFACTOL Q1:
    CTFA Name: Ricinoleamido-propyl Trimonium Chloride
    Pour Point C: -15
    Iodine Value: N/A
    Water Solubility: Soluble

SURFACTOL Q2:
    CTFA Name: Hydroxy Stearamidopropyl Trimonium Chloride
    Melting Point C: 35
    Iodine Value: 5
    Water Solubility: Dispersible

SURFACTOL Q3:
    CTFA Name: Hydroxy Stearamidopropyl Trimonium Methyl Sulfate
    Melting Point C: 45
    Iodine Value: 5
    Water Solubility: Dispersible

CASCHEM INC.: Emulsifiers--SURFACTOL Non-Ionic Surfactants:

The SURFACTOL series of nonionic emulsifiers are ethoxylated castor oils with varying amounts of ethylene oxide added to the hydroxyl bearing fatty acid chain. They vary from self-emulsifiable to completely water soluble. The HLB values vary from 3.6 to 16.

SURFACTOL nonionic surfactants offer these benefits:

* Low odor
* Low foaming
* Excellent stability over broad pH range
* Lubricity
* Excellent fragrance solubilizer

Non-Ionics:

SURFACTOL 318:
    CTFA Name: PEG-5 Castor Oil
    Pour Point C: -25
    Iodine value: 70
    Water Solubility: Dispersible

SURFACTOL 365:
    CTFA Name: PEG-40 Castor Oil
    Pour Point C: 10
    Iodine Value: 36
    Water Solubility: Soluble

NATURECHEM THS-200:
    CTFA Name: PEG-200 Trihydroxystearin
    Melting Point C: 53
    Iodine Value: 5
    Water Solubility: Soluble

CENTRAL SOYA: Lecithins:

CENTROLEX Series:
CENTROLEX is a series of highly concentrated, versatile lecithin powders in an essentially oil-free form:
  CENTROLEX F
  CENTROLEX P

CENTROLENE Series:
CENTROLENE is a series of heavy-bodied lecithins produced through a patented chemical modification known as hydroxylation. This insures a more hydrophilic product. It may also provide some crosslink points in resin-binder systems. CENTROLENE products are more polar than unmodified lecithins.

CENTROPHIL Series:
The CENTROPHIL series is a blend of oil-free phospholipids and refined specialty oils. This combination enhances compatibility with target systems.
CENTROPHIL W is a low viscosity fluid suitable for use as a release or wetting agent.
CENTROPHIL K is a solid, meltable compound high in phospholipids for use in applications where liquid oil is not desirable or viscosity may be a problem.

CENTROPHASE and CENTROPHASE HR:
CENTROPHASE is a series of versatile, low viscosity fluid lecithin products. CENTROPHASE HR is a patented heat resistant lecithin which will retain its color even when exposed to temperatures of 475F. This is important for plastic molding release agents.
Three additional products based on CENTROPHASE HR technology are also available--CENTROPHASE HR2, CENTROPHASE HR3 and CENTROPHASE HR4.

CENTROMIX Series:
The CENTROMIX Series of fluid lecithins is specially formulated to provide optimum pigment dispersion in water-based paints and coatings.

CENTROL Series:
Products in the CENTROL Series are stable, easy-to-handle fluid blends of natural phospholipids and soybean oil.

ACTIFLO:
ACTIFLO is a fluid lecithin rich in phosphatides prepared in a concentrated, undiluted form by a special process.

CENTRAL SOYA: Lecithins(Continued):

Product:

CENTROLEX P:
  Physical State: Granule
  Acetone Insolubles Min. %: 98.0
  Specific Gravity: 1.1

CENTROLEX F:
  Physical State: Coarse Powder
  Acetone Insolubles Min. %: 95.0
  Specific Gravity: 1.1

CENTROLENE A:
  Physical State: Heavy-Bodied Fluid
  Acetone Insolubles Min. %: 58.0
  Specific Gravity: 1.03
  Viscosity Brookfield @ 25C: Max.: 50,000

CENTROLENE S:
  Physical State: Heavy-Bodied Fluid
  Acetone Insolubles Min. %: 58.0
  Specific Gravity: 1.03
  Viscosity Brookfield @ 25C: Max.: 50,000

CENTROPHIL W:
  Physical State: Light Amber Fluid
  Acetone Insolubles Min. %: 35.0
  Specific Gravity: .98
  Viscosity Brookfield @ 25C: Max.: 150

CENTROPHIL CR3:
  Physical State: Amber Fluid
  Acetone Insolubles Min. %: 30.0
  Specific Gravity: .99
  Viscosity Brookfield @ 25C: 300

CENTROPHASE 31:
  Physical State: Clear Amber Fluid
  Acetone Insolubles Min. %: 60.0
  Specific Gravity: 1.03
  Viscosity Brrokfield @ 25C: 8,500

CENTROPHASE C:
  Physical State: Amber Fluid
  Acetone Insolubles Min. %: 50.0
  Specific Gravity: .99
  Viscosity Brookfield @ 25C: 1,500

CENTRAL SOYA: Lecithins(Continued):

Product:

CENTROPHASE 152:
   Physical State: Amber Fluid
   Acetone Insolubles Min. %: 50.0
   Specific Gravity: .99
   Viscosity Brookfield @ 25C Max.: 1,500

CENTROPHASE HR:
   Physical State: Amber Fluid
   Acetone Insolubles Min. %: 50.0
   Specific Gravity: .99
   Viscosity Brookfield @ 25C Max.: 3,000

CENTROPHASE HR2:
   Physical State: Amber Fluid
   Acetone Insolubles Min. %: 60.0
   Specific Gravity: 1.03
   Viscosity Brookfield @ 25C Max.: 8,000

CENTROPHASE HR3:
   Physical State: Amber Fluid
   Acetone Insolubles Min. %: 52.0
   Specific Gravity: .99
   Viscosity Brookfield @ 25C Max.: 2,500

CENTROMIX LP-250:
   Physical State: Amber Fluid
   Acetone Insolubles Min. %: 48.0
   Specific Gravity: 1.01
   Viscosity Brookfield @ 25C Max.: 10,000

CENTROMIX LP-200:
   Physical State: Amber Fluid
   Acetone Insolubles Min. %: 50.0
   Specific Gravity: 1.01
   Viscosity Brookfield @ 25C Max.: 10,000

CENTROL CA:
   Physical State: Amber Fluid
   Acetone Insolubles Min. %: 60.0
   Specific Gravity: 1.03
   Viscosity Brookfield @ 25C Max.: 12,000

CENTROL 2F-UB:
   Physical State: Amber Fluid
   Acetone Insolubles Min. %: 62.0
   Specific Gravity: 1.03
   Viscosity Brookfield @ 25C Max.: 12,000

CENTRAL SOYA: Lecithin(Continued):

Product:

CENTROL 2F-SB:
   Physical State: Amber Fluid
   Acetone Insolubles Min. %: 62.0
   Specific Gravity: 1.03
   Viscosity Brookfield @ 25C Max: 12,000

CENTROL 2F-DB:
   Physical State: Amber Fluid
   Acetone Insolubles Min. %: 62.0
   Specific Gravity: 1.03
   Viscosity Brookfield @ 25C Max: 12,000

CENTROL 3F-UB:
   Physical State: Amber Fluid
   Acetone Insolubles Min. %: 62.0
   Specific Gravity: 1.03
   Viscosity Brookfield @ 25C Max: 12,000

CENTROL 3F-SB:
   Physical State: Amber Fluid
   Acetone Insolubles Min. %: 62.0
   Specific Gravity: 1.03
   Viscosity Brookfield @ 25C Max.: 12,000

CENTROL 3F-DB:
   Physical State: Amber Fluid
   Acetone Insoubles Min. %: 62.0
   Specific Gravity: 1.03
   Viscosity Brookfield @ 25C Max.: 12,000

ACTIFLO 68-UB:
   Physical State: Heavy-Bodied Fluid
   Acetone Insolubles Min. %: 66.0
   Specific Gravity: 1.04
   Viscosity Brookfield @ 25C Max.: 15,000

ACTIFLO 68-SB:
   Physical State: Heavy-Bodied Fluid
   Acetone Insolubles Min. %: 66.0
   Specific Gravity: 1.04
   Viscosity Brookfield @ 25C Max.: 15,000

ACTIFLO 68-DB:
   Physical State: Heavy-Bodied Fluid
   Acetone Insolubles Min. %: 66.0
   Specific Gravity: 1.04
   Viscosity Brookfield @ 25C Max.: 15,000

W.A. CLEARY PRODUCTS, INC.: CLEARATE Lecithin:

Lecithin is a surface active agent which reduces the surface tension of vegetable oils and in percentages less than 1% will form water in oil emulsions. And at higher rates will form oil in water emulsions. Whenever it is desired to incorporate lecithin direct into the water phase it is more desirable to use a water dispersible type such as CLEARATE WDF.

CLEARATE F:
This special lecithin is superior to CLEARATE lecithin for water in oil emulsions.

CLEARATE WDF:
Designed for oil in water emulsions.

WDF 2:
Designed for oil in water emulsions but exhibiting superior qualities to WDF, and is slightly more expensive.

Note: In forming emulsions, best results are obtained when the CLEARATES are added to oil instead of to water.

LECITHIN CLEARATES
Food Applications for Lecithins

CLEARATES:

B60:Unbleached Pourable
 Acetone Insoluble: 62-65
 Viscosity @ 77F in seconds Gardner Holdt: 120 max.

B70:Single Bleached Pourable
 Acetone Insoluble: 62-65
 Viscosity @ 77F in seconds Gardner Holdt: 120 max.

B70L:Double Bleached Pourable
 Acetone Insoluble: 60-65
 Viscosity @ 77F in seconds Gardner Holdt: 60-90 max.

WDF:Water Dispersible Food Grade
 Acetone Insoluble: 56-58
 Viscosity @ 77F in seconds Gardner Holdt: 60-90 max.

HYDROXYLATED:
 Acetone Insoluble: 58-60
 Viscosity @ 77F in seconds Gardner Holdt: 60-80 max.

CLEARATE-Q:
 Acetone Insoluble: 62-65
 Viscosity @ 77F in seconds Gardner Holdt: 60-90 max.

CLEARATE-F:
 Acetone Insoluble: 56-58
 Viscosity @ 77F in seconds Gardner Holdt: 50-60 max.

CLIMAX PERFORMANCE MATERIALS CORP.: ACTRABASE Emulsifiers:

Metalworking fluid emulsifier bases for soluble oils and semi-synthetic fluids.

ACTRABASE:

ACTRABASE 31-A:
  Premium quality base for naphthenic oils.

ACTRABASE 215:
  Low use level base for paraffinic oils.

ACTRABASE 264:
  Low use level for naphthenic oils.

ACTRABASE PS-470:
  Medium molecular weight petroleum sulfonate, sodium salt.

ACTRABASE 1963:
  Medium molecular weight petroleum sulfonate, sodium salt/ oil blend.

ACTRABASE SS-503:
  Semi-synthetic concentrate base for total fluid.

ACTRABASE SS-523:
  SS-503 with no oil or water in concentrate. Designed to add naphthenic oil and water for total fluid.

ACTRASOL Emulsifiers & Lubricants:

Sulfated oils and fatty acids for use as lubricants and emulsifiers in metalworking fluids.

ACTRASOL C-75:
  Sulfated castor oil

ACTRASOL PSR:
  Sulfated ricinoleic acid

ACTRASOL MY-75:
  Sulfated soya ester

CYCLO CHEMICALS CORP.: CYCLOCHEM:

Fatty Acid Esters:

Excellent for a variety of cosmetic and industrial formulators, as emollients, emulsifiers, thickeners, stabilizers, opacifiers, pearlizers, dispersants. A number of customized fatty acid blends for specialized applications are also available. In addition, Cyclo will custom tailor fatty acid esters for the formulator's specific requirements.

Product:

CYCLOCHEM EGDS:
    CTFA Nomenclature: Glycol Distearate
    Physical Form: Off-white flakes
    Features: Viscosifier, Opacifier
    Applications: Shampoos, liquid detergents, lotions

CYCLOCHEM EGMS:
    CTFA Nomenclature: Glycol Stearate
    Physical Form: White beads, flakes
    Features: Viscosifier, Pearlizer
    Applications: Shampoos, lotions, detergents where bright pearl is desired.

CYCLOCHEM EGMS/SE:
    CTFA Nomenclature: Glycol Stearate SE
    Physical Form: Off-white flakes
    Features: Self-emulsifying
    Applications: Emulsifier for O/W lotions and cremes.

CYCLOCHEM GMS:
    CTFA Nomenclature: Glyceryl Stearate
    Physical Form: White beads, flakes
    Features: Emulsifier, emollient
    Applications: Lipophilic emulsifier for cremes/lotions

CYCLOCHEM GMS 165:
    CTFA Nomenclature: Glyceryl Stearate (and) PEG 100 Stearate
    Physical Form: Off-white flakes
    Features: Acid stable, Self-emulsifying
    Applications: Nonionic modified GMS for acid stable applications.

CYCLOCHEM LVL:
    CTFA Nomenclature: Lauryl Lactate
    Physical Form: Off-white liquid
    Features: Penetrant, emollient, dispersant
    Applications: Antiperspirants, powders, lotions, cremes where non oily penetration is desired

CYCLO CORP.: CYCLOCHEM(Continued):

Product:

CYCLOCHEM MM/M:
  CTFA Nomenclature: Myristyl Myristate
  Physical Form: Off-white flakes
  Features: Light emollient
  Applications: Lotions, cremes where velvety after-feel desired

CYCLOCHEM PETS:
  CTFA Nomenclature: Pentaerythritol Tetrastearate
  Physical Form: Off-white beads, flakes
  Features: Synthetic Wax Lubricant
  Applications: Polishes, coatings, hot melts

CYCLOCHEM SPS:
  CTFA Nomenclature: Cetyl Esters
  Physical Form: Off-white flakes
  Features: Synthetic Spermaceti
  Applications: Cremes, sticks, aerosols, meets NF

CYCLOCHEM SS:
  CTFA Nomenclature: Stearyl Stearate
  Physical Form: Off-white flakes
  Features: Viscosifying emollient
  Applications: Firming agent for lotions, cremes, polishes

Polyethylene Glycol Ester Group:

CYCLOCHEM PEG:
  Molecular Weight Range: 200, 300, 400, 600, 6000
  Esters: Mono & Diester laurates, stearates & oleates
  Features: Dispersing and thickening in both water and oil systems
  Applications: Solid & liquid emulsifiers, emollients and conditioners

Blended Products:

CYCLOCHEM EM324:
  Major Components: Natural fatty acids & alcohols
  Physical Form: Off-white flakes
  Features: Self-emulsifying wax
  Applications: o/w cremes, lotions

CYCLOCHEM 326A:
  Major Components: Natural fatty acids & alcohols
  Physical Form: Off-white flakes
  Features: CTFA: Synthetic Beeswax
  Applications: w/o cremes, cold cremes, lipsticks

CYCLO CORP.: CYCLOCHEM(Continued):

Blended Products:

Product:

CYCLOCHEM EM560:
    CTFA Nomenclature: Natural fatty acids & alcohols
    Physical Form: Off-white solid
    Features: Self-emulsifying wax
    Applications: o/w cremes and lotion where good tolerance for peroxide is required.

CYCLOCHEM POL:
    CTFA Nomenclature: Higher fatty alcohols & ethoxylates
    Physical Form: Off-white solid
    Features: Emulsifier for alkaline systems
    Applications: o/w cremes, lotions, relaxers containing caustic

CYCLOSHEEN 202:
    CTFA Nomenclature: EGMS, Emulsifiers
    Physical Form: Soft paste
    Features: Room temperature dispersible
    Applications: Pearlizing agent for shampoos

Other Available Products:
CYCLOCHEM CL:
    CTFA Nomenclature: Cetyl Lactate
CYCLOCHEM CP:
    CTFA Nomenclature: Cetyl Palmitate
CYCLOCHEM GTS:
    CTFA Nomenclature: Tristearin
CYCLOCHEM GTL:
    CTFA Nomenclature: Trilaurin
CYCLOCHEM GMO:
    CTFA Nomenclature: Glyceryl Oleate
CYCLOCHEM GTO:
    CTFA Nomenclature: Triolein
CYCLOCHEM GTIS:
    CTFA Nomenclature: Triisostearin
CYCLOCHEM INEO:
    CTFA Nomenclature: Isostearyl Neopentanoate
CYCLOCHEM IPM:
    CTFA Nomenclature: Isopropyl Myristate
CYCLOCHEM IPP:
    CTFA Nomenclature: Isopropyl Palmitate
CYCLOCHEM ML:
    CTFA Nomenclature: Myristyl Lactate
CYCLOCHEM MST:
    CTFA Nomenclature: Myristyl Stearate
CYCLOCHEM PETO:
    CTFA Nomenclature: Pentaerythritol Tetraoleate
CYCLOCHEM PGMS:
    CTFA Nomenclature: Propylene Glycol Stearate

DESOTO, INC.: Esters & Ethoxylated Esters:

Product Name:

DESOTAN SMO:
  % Active: 100
  Form: Liquid

DESOTAN SMT:
  % Active: 100
  Form: Liquid

Application: Lipophilic emulsifiers, fiber lubricants and softeners. Insoluble in water, but soluble in oils and organic solvents.

DESOTAN SMO-20:
  % Active: 100
  n: 20
  Cloud Point, F: 149-158   10% NaCl
  Form: Liquid

DESOTAN SMT-20:
  % Active: 100
  n: 20
  Cloud Point, F: 149-158   10% NaCl
  Form: Liquid

Application: Hydrophilic emulsifiers and wetting agents. General purpose nonionic surfactants exhibiting antistatic and lubricating properties.

DESOTO, INC.: Ethoxylated Amines:

DESOMEEN TA-2:
  % Active: 99
  n: 2
  Form: Paste

DESOMEEN TA-15:
  % Active: 100
  n: 15
  Cloud Point, F: 172-179
  Form: Liquid

DESOMEEN TA-20:
  % Active: 100
  n: 20
  Cloud Point, F: 179-181   10% NaCl
  Form: Liquid

Application: Emulsifiers and dispersants. Used as textile scouring agents, textile dyeing assistants, desizing assistants, softening agents, antistatic agents, etc.

DESOTO, INC.: Emulsifying Agents:

Octyl Phenol Ethoxylates:

DESONIC S-45:
   % Active: 100
   n: 4.5
   Form: Liquid
   Application: Emulsifier, Oil Soluble Surfactant, Emulsion Cleaner, Dry Dishwashing Detergent, Polish Emulsifier.
DESONIC S-100:
   % Active: 100
   n: 9-10
   Cloud Point, F: 140-158
   Form: Liquid
   Application: Metal & Textile Processing, Household & Industrial Cleaners, Emulsifier for Vinyl & Acrylic Polymerization.
DESONIC S-405:
   % Active: 70
   n: 40
   Cloud Point, F: 165-176
                  10% NaCl
   Form: Liquid
   Application: Co-Emulsifier for Vinyl & Acrylic Polymerization, Dye Assistant

Linear Alcohol Ethoxylates:

DESONIC 3K:
   % Active: 100
   n: 3
   Form: Liquid
   Application: Detergent, Emuulsifier, Wetting, Oil Soluble Surfactant, Defoamer
DESONIC 5K:
   % Active: 100
   n: 5
   Cloud Point, F: 75-79
   Form: Liquid
   Application: Detergent, Emulsifier, Wetting for Household and Industrial Detergents
DESONIC 7K:
   % Active: 100
   n: 7
   Cloud Point, F: 118-122
   Form: Liquid
   Application: Detergent, Emulsifier, Wetting for Household and Industrial Detergents
DESONIC 9K:
   % Active: 100
   n: 9
   Cloud Point, F: 131-138
   Form: Liquid
   Application: Detergent, Emulsifier, Wetting for Household and Industrial Detergents

DESOTO, INC.: Phosphate Esters:

Product Name:

DESOPHOS 4 NP:
 % Active: 100
 Form: Liquid
 Application: Superior emulsifier, useful in emulsion polymerization. Most oil-soluble member of DESOPHOS NP series. Water soluble when neutralized.

DESOPHOS 6 NP:
 % Active: 100
 Form: Liquid
 Application: Excellent emulsifying and detergent properties. Used in emulsion polymerization.

DESOPHOS 9 NP:
 % Active: 100
 Form: Liquid
 Application: Dedusting agent for dry cleaning detergent, alkaline powders, water-repellent fabric finishes. Used in emulsion polymerization for formation of polyvinyl acetate and acrylic films.

DESOPHOS 30 NP:
 % Active: 100
 Form: Liquid
 Application: Emulsifier and stabilizing agent. Useful in preparation of polyvinyl acetate and acrylic copolymers.

DESOPHOS 6 NP4:
 % Active: 100
 Form: Liquid
 Application: High electrolyte tolerance. May be used in waterless hand cleaners and laundry detergents. Polymerization emulsifier for polyvinyl acetate and acrylic films.

DESOPOHOS 4 CP:
 % Active: 95
 Form: Liquid
 Application: Used as a softener and antistatic agent in textile finishing. Useful in lubricants for filament yarns, synthetic fibers, and wool. May be used to emulsify cosmetic oils and creams, and for the polymerization of latices.

DESOPHOS 7 OPNa:
 % Active: 98
 Form: Liquid
 Application: Used as a dispersible surfactant in the emulsification of mineral oils. A softener and antistatic agent for textiles. Soluble in aromatic solvents. Used for formulating lubricants for synthetic and wool fibers.

DESOTO, INC.: Phosphate Esters(Continued):

Product Name:

DESOPHOS 3 OMP:
    % Active: 100
    Form: Liquid
    Application: Acid ester of an aliphatic hydrophobic base. Outstanding emulsifier for paraffinic oils.
DESOPHOS 6 DNP:
    % Active: 100
    Form: Liquid
    Application: Superior paraffinic oil emulsifier as triethanolamine salt. Imparts rust-inhibiting properties. Most oil-soluble member of DESOPHOS DNP series.
DESOPHOS 8 DNP:
    % Active: 100
    Form: Liquid
    Application: Dispersible in water. Soluble in aromatic solvents and Kerosene. Used in industrial cleaners and in dry cleaning. Spontaneously emulsifies phosphated and chlorinated pesticide concentrates and polyethylene.
DESOPHOS 14 DNP:
    % Active: 100
    Form: Liquid
    Application: Water soluble. Similar to DESOPHOS 8 DNP.

DESOTO, INC.: Castor Oil Ethoxylates:

Product Name:

DESONIC 30C:
    % Active: 100
    n: 30
    Form: Liquid
DESONIC 36C:
    % Active: 100
    n: 36
    Cloud Point, F: 122-140
    Form: Liquid
DESONIC 40C:
    % Active: 100
    n: 40
    Cloud Point, F: 173-179
    Form: Liquid
DESONIC 54C:
    % Active: 100
    n: 54
    Cloud Point, F: 136-142 10% NaCl
    Form: Liquid

Application: Emulsifiers, Lubricants, Dye Levelors, Antistatic Agents, and Dispersants for Textiles. Emulsifying Agents for Polyurethane Foams; Softening and Rewetting Agents for Wet Strength Paper.

DUPONT CO.: DUPONOL Alcohol Sulfates as Emulsifying Agents:

The DUPONOL alcohol sulfate surface active agents are products possessing excellent emulsifying and dispersing characteristics. They are employed for the preparation of stable emulsions of solvents and oils for use in scouring and finishing textiles, for fatliquoring leather and degreasing skins, for furniture and automobile cleaners and for paints. They are also suitable for the preparation of emulsions of waxes and resins for polishing and finishing compositions, for textile sizes, for pharmaceutical and cosmetic emulsions and for many other purposes. Emulsions prepared with suitable DUPONOL surface active agents are quite resistant to the action of acids, alkalies and electrolytes.

The DUPONOL surface active agents are also employed as dispersing agents in the preparation of a variety of emulsions of limited stability. Emulsions of this type find application in agricultural, insecticidal, pharmaceutical and dust-laying sprays.

Because of the high emulsifying power of the DUPONOL surface active agents, clear, transparent emulsions can be prepared with them. These emulsions are of the O/W type which on casual examination appear to be true solutions. Clear emulsions have a decided sales appeal and to obtain such emulsions it is necessary to use fairly definite ratios of oil, water and the appropriate DUPONOL product.

Product:

DUPONOL SP:
   Major surfactant component: Sodium octyl sulfate
   Active Ingredient, %: 33-35
   Physical Form: Pale yellow liquid
   Odor: Bland

DUPONOL WN:
   Major surfactant component: Sodium octyl/decyl sulfate
   Active Ingredient, %: 33-35
   Physical Form: Pale yellow liquid
   Odor: Bland

DUPONOL WAQ:
   Major surfactant component: Sodium lauryl sulfate
   Active Ingredient, %: 28-30
   Physical Form: Pale yellow, noticeably viscous fluid
   Odor: Bland

DUPONOL WAQE:
   Major surfactant component: Sodium lauryl sulfate
   Active Ingredient, %: 29-31
   Physical Form: Slightly viscous, pale yellow liquid
   Odor: Bland

DUPONT CO.: DUPONOL Alcohol Sulfates as Emulsifying Agents (Continued):

Product:

DUPONOL QC:
   Major surfactant component: Sodium lauryl sulfate
   Active Ingredient, %: 29-31
   Physical Form: Slightly viscous, almost colorless liquid
   Odor: Bland

DUPONOL WA Paste:
   Major surfactant component: Sodium lauryl sulfate
   Active Ingredient, %: 29-31
   Physical Form: Pale yellow viscous liquid
   Odor: Bland

DUPONOL ME Dry:
   Major surfactant component: Sodium lauryl sulfate
   Active Ingredient, %: 90-96
   Physical Form: White to cream-colored powder
   Odor: Mild fatty

DUPONOL C:
   Major surfactant component: Sodium lauryl sulfate
   Active Ingredient, %: 90-96
   Physical Form: White to cream-colored powder
   Odor: Mild fatty

DUPONOL D:
   Major surfactant component: Sodium lauryl/oleyl sulfate
   Active Ingredient, %: 38
   Physical Form: Yellow to pale brown viscous paste
   Odor: Mild fatty

DUPONOL LS:
   Major surfactant component: Sodium oleyl/lauryl sulfate
   Active Ingredient, %: 26
   Physical Form: Yellow to tan fluid paste
   Odor: Mild fatty

DUPONT CO.: DUPONOL Alcohol Sulfates as Emulsifying Agents (Continued):

Product:

DUPONOL G:
  Major Surfactant Component: Amine Salt of lauryl sulfate
  Active Ingredient, %: 92
  Physical Form: Yellow to light amber viscous paste
  Odor: Mild fatty

DUPONOL EP:
  Major Surfactant Component: Diethanolamine salt of lauryl sulfate
  Active Ingredient, %: 33-36
  Physical Form: Pale yellow slightly viscous paste
  Odor: Bland

DUPONOL RA:
  Major Surfactant Component: Sodium alkyl ether sulfate
  Active Ingredient, %: 34
  Physical Form: Light amber liquid
  Odor: Bland

DUPONOL XL:
  Major Surfactant Component: Amine Salt of lauryl sulfate plus amphoteric salt
  Active Ingredient, %: 36
  Physical Form: Yellow, moderately viscous liquid
  Odor: Faintly spicy

DUPONOL FAS:
  Major Surfactant Component: Ethoxylated alkyl sulfate plus nonionic surfactant
  Active Ingredient, %: 50
  Physical Form: Pale yellow liquid
  Odor: Alcoholic

AVITEX AD:
  Major Surfactant Component: Sodium salt of sulfated oleyl acetate
  Active Ingredient, %: 50
  Physical Form: Red-brown liquid
  Odor: Sweet Alcoholic

DUPONT CO.: Surfactants: Wetting Agents/Dispersing Agents/ Emulsifying Agents:

Anionic:
Alcohol Sulfates, Sodium Salts:

AVITEX AD:
Active Ingredient %: 30

DUPONOL C (NF):
Active Ingredient %: 95

DUPONOL D (Paste):
Active Ingredient %: 38

DUPONOL FAS:
Active Ingredient %: 50

DUPONOL KW (Dry):
Active Ingredient %: 60

DUPONOL LS (Paste):
Active Ingredient %: 26

DUPONOL ME (Dry):
Active Ingredient %: 95

DUPONOL QC:
Active Ingredient %: 30

DUPONOL RA:
Active Ingredient %: 34

DUPONOL SP:
Active Ingredient %: 35

DUPONOL WAQ, WAQE:
Active Ingredient %: 30

DUPONOL WA (Paste):
Active Ingredient %: 30

DUPONOL WN:
Active Ingredient %: 35

Alcohol Sulfates, Amine Salts:

DUPONOL EP:
Active Ingredient %: 35

DUPONOL G:
Active Ingredient %: 95

DUPONOL XL:
Active Ingredient %: 35

DUPONT CO.: DUPONT Surfactants: Wetting Agents/Dispersing Agents/
   Emulsifying Agents(Continued):

<u>Anionic:</u>
<u>Alcohol Phosphates:</u>

ZELEC NE:
   Active Ingredient %: 100

ZELEC NK:
   Active Ingredient %: 100

ZELEC TY:
   Active Ingredient %: 50

ZELEC UN:
   Active Ingredient %: 100

Sulphonates, Aliphatic:

ALKANOL 189-S:
   Active Ingredient %: 30

PETROWET R:
   Active Ingredient %: 24

Sulfonates, Alkylaryl:

ALKANOL ND:
   Active Ingredient %: 45

ALKANOL XC:
   Active Ingredient %: 100

ALKANOL WXN:
   Active Ingredient %: 30

DUPONT CO.: Surfactants: Wetting Agents/Dispersing Agents/
Emulsifying Agents(Continued):

Nonionic:
Alcohol/Ethylene Oxide Adducts:

MERPOL 100(HLB=13.5):
    Active Ingredient %: 100

MERPOL HCS(HLB=15.3):
    Active Ingredient %: 60

MERPOL OJ(HLB=12.5):
    Active Ingredient %: 100

MERPOL SE(HLB=10.5):
    Active Ingredient %: 95

MERPOL SH(HLB=12.8):
    Active Ingredient %: 50

Specialties:

ALKANOL A-CN:
    Active Ingredient %: 60

MERPOL A:
    Active Ingredient %: 100

MERPOL DA:
    Active Ingredient %: 60

MERPOL LF-H:
    Active Ingredient %: 50

ALKANOL 6112:
    Active Ingredient %: 50

**DUPONT CO.: Surfactants: Wetting Agents/Dispersing Agents/
Emulsifying Agents(Continued):**

Fluorosurfactants:
Anionic:

ZONYL FSA:
    Active Ingredient %: 50

ZONYL FSE:
    Active Ingredient %: 14

ZONYL FSJ:
    Active Ingredient %: 40

ZONYL FSP:
    Active Ingredient %: 35

ZONYL TBS:
    Active Ingredient %: 33

ZONYL UR:
    Active Ingredient %: 100

Cationic:

ZONYL FSC:
    Active Ingredient %: 50

Nonionic:

ZONYL FSN:
    Active Ingredient %: 50

ZONYL FSN-100:
    Active Ingredient %: 100

ZONYL FSO:
    Active Ingredient %: 50

ZONYL FSO-100:
    Active Ingredient %: 100

Amphoteric:

ZONYL FSK:
    Active Ingredient %: 47

EASTMAN CHEMICAL PRODUCTS, INC.: Food-Grade Emulsifiers:

MYVEROL Distilled Monoglycerides:

Type:

18-00:
    Fat Source: Hydrogenated lard or tallow
    Monoester Content Min %: 90
    Melting Point C (F): 68 (154)
18-04K:
    Fat Source: Hydrogenated palm oil or palm stearine
    Monoester Content Min %: 90
    Melting Point C (F): 66 (151)
18-06(K):
    Fat Source: Hydrogenated soybean oil
    Monoester Content Min %: 90
    Melting Point C (F): 69 (156)
18-07K:
    Fat Source: Hydrogenated vegetable oil or cottonseed oil
    Monoester Content Min %: 90
    Melting Point C (F): 68 (154)
18-30:
    Fat Source: Edible beef tallow
    Monoester Content Min %: 90
    Melting Point C (F): 60 (140)
18-35K:
    Fat Source: Refined palm oil
    Monoester Content Min %: 90
    Melting Point C (F): 60 (140)
18-40:
    Fat Source: Edible lard
    Monoester Content Min %: 90
    Melting Point C (F): 58 (136)
18-50K:
    Fat Source: Partially hydrogenated soybean oil
    Monoester Content Min %: 90
    Melting Point C (F): 54 (129)
18-85K:
    Fat Source: Cottonseed oil
    Monoester Content Min %: 90
    Melting Point C (F): 46 (115)
18-92(K):
    Fat Source: Sunflower oil
    Monoester Content Min %: 90
    Melting Point C (F): 41 (106)
18-99K:
    Fat Source: Low-erucic rapeseed oil
    Monoester Content Min %: 90
    Melting Point C (F): 35 (94)

EASTMAN CHEMICAL PRODUCTS, INC.: Food-Grade Emulsifiers (Continued):

MYVACET Distilled Acetylated Monoglycerides:

Type:

5-07(K):
   Fat Source: Hydrogenated vegetable oil
   Form: Waxy solid
   Hydroxyl Value: 133-152
   Melting Point C(F): 41-46 (105-114)
   Saponification Value: 279-292

7-00:
   Fat Source: Hydrogenated lard
   Form: Waxy solid
   Hydroxyl Value: 80.5-95
   Melting Point C(F): 37-40 (99-104)
   Saponification Value: 316-331

7-07K:
   Fat Source: Hydrogenated vegetable oil
   Form: Waxy solid
   Hydroxyl Value: 80.5-95
   Melting Point C(F): 37-40 (99-104)
   Saponification Value: 316-331

9-08K:
   Fat Source: Hydrogenated coconut oil
   Form: Liquid
   Hydroxyl Value: 0-20
   Melting Point C(F): 4-12 (40-54)
   Saponification Value: 410-440

9-40:
   Fat Source: Edible lard
   Form: Liquid
   Hydroxyl Value: 0-15
   Melting Point C(F): 4-12 (40-54)
   Saponification Value: 375-385

9-45K:
   Fat Source: Partially hydrogenated soybean oil
   Form: Liquid
   Hydroxyl Value: 0-15
   Melting Point C(F): 4-12 (40-54)
   Saponification Value: 370-382

EASTMAN CHEMICAL PRODUCTS, INC.: Food-Grade Emulsifiers (Continued):

MYVAPLEX Concentrated Glyceryl Monostearates:

Type:

600(K):
   Fat Source: Hydrogenated soybean oil
   Form: Small beads
   Monoester Content Min %: 90
   Melting Point C (F): 69 (156)
600P(K):
   Fat Source: Hydrogenated soybean oil
   Form: Powder
   Monoester Content Min %: 90
   Melting Point C (F): 69 (156)

MYVATEM Dispersing Agents:

Type:

06K:
   Fat Source: Hydrogenated soybean oil
   Melting Point C (F): 47 (117)
   Specifications: Conforms to Food Chemicals Codex III, pp.98-99
30:
   Fat Source: Edible tallow
   Melting Point C (F): 33 (91)
   Specifications: Conforms to Food Chemicals Codex III, pp.98-99
35K:
   Fat Source: Refined palm oil
   Melting Point C (F): 26 (79)
   Specifications: Conforms to Food Chemicals Codex III, pp.98-99
92K:
   Fat Source: Refined sunflower oil
   Melting Point C (F): <0 (<32)
   Specifications: Conforms to Food Chemicals Codex III, pp.98-99

MYVEROL P-06 Distilled Propylene Glycol Monoester:

P-06(K):
   Fat Source: Hydrogenated soybean oil
   Monoester Content Min %: 90
   Melting Point C (F): 45 (113)

MYVEROL SMG Succinylated Monoglycerides:

SMG VK:
   Fat Source: Hydrogenated palm oil or palm stearine
   Monoester Content %: 12-20
   Melting Point C (F): 58 (136)

EASTMAN CHEMICAL PRODUCTS, INC.: MYVATEX Food Emulsifier Blends:

Type:

3-50K:
  Fat Source: Soybean oil
  Form: Beads
  Composition: Distilled monoglycerides and distilled propylene glycol monoesters
    Monoester Content Min %: 90
    Melting Point C (F): 58 (136)

7-85K:
  Fat Source: Cottonseed oil
  Form: Plastic
  Composition: Distilled monoglycerides with 30% cottonseed oil
    Monoester Content Min %: 63
    Melting Point C (F): 49 (120)

8-06K:
  Fat Source: Soybean oil, sunflower oil
  Form: Beads
  Composition: Distilled monoglycerides with 20% hydrogenated soybean oil
    Monoester Content Min %: 72
    Melting Point C (F): 67 (153)

8-16K:
  Fat Source: Palm oil
  Form: Beads
  Composition: Distilled monglycerides with 20% hydrogenated palm oil
    Monoester Content Min %: 72
    Melting Point C (F): 61 (142)

8-20:
  Fat Source: Lard, tallow, soybean oil
  Form: Beads
  Composition: Distilled monglycerides with 20% hydrogenated soybean oil
    Monoester Content Min %: 72
    Melting Point C (F): 58 (136)

8-20E:
  Fat Source: Lard, tallow, sunflower oil, soybean oil
  Form: Beads or powder
  Composition: Distilled monoglycerides with 20% hydrogenated soybean oil
    Monoester Content Min %: 72
    Melting Point C (F): 57 (135)

EASTMAN CHEMICAL PRODUCTS, INC.: MYVATEX Food Emulsifier Blends (Continued):

Type:

25-07(K):
  Fat Source: Soybean oil
  Form: Soft plastic
  Composition: Distilled monoglycerides, lecithin, water, propionic acid, and sodium propionate
  Monoester Content Min %: 25

40-06S:
  Fat Source: Soybean oil
  Form: Soft plastic
  Composition: Distilled propylene monesters, distilled monoglycerides, lactylic esters of fatty acids (stearic), water, and potassium sorbate
  Monoester Content Min %: 25

90-10K:
  Fat Source: Rapeseed oil, cottonseed oil
  Form: Beads
  Composition: Hydrogenated rapeseed oil, hydrogenated cottonseed oil
  Monoester Content Min %: 0
  Melting Point C (F): 61 (142)

DO CONTROL Strengthener(K):
  Fat Source: Palm oil
  Form: Powder
  Composition: Distilled succinylated monglycerides and distilled monoglycerides
  Monoester Content Min %: 41
  Melting Point C (F): 53 (127)

LIQUID LITE Food Emulsifier(K):
  Fat Source: Soybean oil, cottonseed oil
  Form: Beads
  Composition: Distilled propylene glycol monoesters and distilled acetylated monoglycerides
  Monoester Content Min %: 80
  Melting Point C (F): 44-48 (111-118)

MIGHTY SOFT Softener(K):
  Fat Source: Soybean oil
  Form: Powder
  Composition: Distilled monoglycerides
  Monoester Content Min %: 90
  Melting Point C (F): 67 (153)

EASTMAN CHEMICAL PRODUCTS, INC.: MYVATEX Food Emulsifier Blends (Continued):

Type:

MONOSET Food Emulsifier K:
   Fat Source: Rapeseed oil, cottonseed oil, palm oil
   Form: Beads
   Composition: Distilled monoglycerides, rapeseed and cottonseed oils
     Monoester Content Min %: 18
     Melting Point C (F): 63 (146)

MSPS:
   Fat Source: Soybean oil
   Form: Beads
   Composition: Distilled monoglycerides with 25% polysorbate 80
     Monoester Content Min %: 67.5
     Melting Point C (F): 69 (156)

SMG 30K:
   Fat Source: Palm oil, soybean oil
   Form: Soft plastic
   Composition: Distilled succinylated monoglycerides, distilled monoglycerides, lecithin, water, and propionic acid
     Monoester Content Min %: 30

SSH:
   Fat Source: Soybean oil
   Form: Soft plastic
   Composition: Distilled monoglycerides, lecithin, water, and propionic acid
     Monoester Content Min %: 45

TEXTURE LITE Food Emulsifier(K):
   Fat Source: Soybean oil
   Form: Powder
   Composition: Distilled monoglycerides, distilled propylene glycol monoesters, sodium stearoyl lactylate with silicon dioxide
     Monoester Content Min %: 80

FMC CORP.: Carrageenan Stabilizer/Emulsifier:

Carrageenan is a generic term applied to hydrocolloids extracted from a number of closely related species of red seaweed.

Chondrus Crispus:
- Also known as Irish Moss
- Small bushy plant about 10 cm in height
- Harvested along coast of the North Atlantic
- Most familiar of the red seaweeds
- Yields kappa and lambda carrageenans

Eucheuma Species:
- Cultivation pioneered by Marine Colloids
- Spiny, bushy plant about 50 cm in height
- Harvested from coral reefs in the Far East
- Yields kappa and iota carrageenans

Gigartina Species:
- Grows worldwide in cold coastal waters
- Bushy or leafy plant ranging up to 5 m in height
- Yields more carrageenan than any other species
- Yields kappa and lambda carrageenans

Type A Products - Carrageenan
Type B Products - Carrageenan with Sugars
Type C Products - Carrageenan, Salts
Type D Products - Carrageenan plus Chemical Additives
Type E Products - Seaweed Flour & Seaweed Flour with Additives
Type F Products - Seaweed Flour with Additives
Type G Products - Poligeenan
Type H Products - Locust Bean Gum
Type I Products - Nutricol Konjac Flour with Additives

GOLDSCHMIDT CHEMICAL CORP.: Emulsifiers for the Preparation of Acid-, Alkaline- and Salt-stable Emulsions:

Product:

TEGINACID, se:
  Glycerol mono distearates mixed with other nonionics
  Appearance/Colour: Powder/ivory
  Total Monoester: 50-60%
  HLB: 12
  Rise Melting Point C: 58-63

TEGINACID X, se:
  Glycerol mono distearates mixed with other nonionics
  Appearance/Colour: Powder/ivory
  Total Monoester: 45-55%
  HLB: 12
  Rise Melting Point C: 55-61

TEGINACID H, se:
  Glycerol mono distearates mixed with other nonionics
  Appearance/Colour: Powder/ivory
  Total Monoester: 20%
  HLB: 11
  Rise Melting Point C: 45-51

TEGINACID KL149, se:
  Glycerol mono distearates mixed with other nonionics
  Appearance/Colour: Powder/ivory
  Total Monoester: 30%
  HLB: 11
  Rise Melting Point C: 60

TEGINACID ML, se:
  Glycerol mono distearates mixed with other nonionics
  Appearance/Colour: Powder/ivory
  Total Monoester: 30%
  HLB: 11
  Rise Melting Point C: 52-58

Remarks: Principally used for acid and salt-resisting emulsions of O/W type of liquid or creamy consistency.
  Use: cosmetics, etc.

GOLDSCHMIDT CHEMICAL CORP.: Emulsifiers for the Preparation of Acid-, Alkaline- and Salt-stable Emulsions(Continued):

Product:

TEGINACID Special, se:
   Glycerol mono distearates mixed with fatty alcohol sulfates
   Appearance/Colour: Powder/white/ivory
   Total Monoester: 35-45%
   HLB: 12
   Rise Melting Point C: 53-59

TEGINACID R, se:
   Glycerol mono distearates mixed with cationic co-emulsifiers
   Appearance/Colour: Powder/white/ivory
   Total Monoester: 45-55%
   HLB: 12
   Rise Melting Point C: 56-60

Remarks: Cationic emulsifier for the manufacture of hair preparations, deodorant preparations, etc.

EMULGATOR E2149, se:
   Blends of nonionics
   Appearance/Colour: waxy/white/ivory
   HLB: 11
   Rise Melting Point C: 41-47

EMULGATOR E2155, se:
   Blends of nonionics
   Appearance/Colour: waxy/white/ivory
   HLB: 11
   Rise Melting Point C: 49-55

EMULGATOR E2209, se:
   Blends of nonionics
   Appearance/Colour: waxy/white/ivory
   HLB: 10
   Rise Melting Point C: 47-53

EMULGATOR E2568, se:
   Blends of nonionics
   Appearance/colour: waxy/white/ivory
   HLB: 16
   Rise Melting Point C: 44-49

Remarks: Emulsifiers and stabilizers for cosmetic and pharmaceutical emulsions of O/W type.
Resistant to alkaline and acid substances.

GOLDSCHMIDT CHEMICAL CORP.: Products for Pharmaceutical Preparations: Emulsifiers:

Product:

TEGIN 4433:
   Glycerol monostearate, se
   Appearance/Colour: Powder/white/ivory
   Total Monoester: 30-40%
   HLB: 12
   Rise Melting Point Value: 54-57
   Remarks: British Pharmaceutical Codex

TEGIN 4011:
   Glycerol monostearate, nse
   Appearance/Colour: Powder/white/ivory
   Total Monoester: ca. 40%
   HLB: 3,8
   Rise Melting Point Value: 54-60
   Remarks: Europ. Arzneibuch

TEGIN A 412:
   Ethylene glycol monostearate, nse
   Appearance/Colour: Powder/white/ivory
   Total Monoester: min. 50%
   HLB: 3,2
   Rise Melting Point Value: 54-57
   Remarks: Pharmacopee Francaise

TEGIN A 422:
   Diethylene glycol monostearate, nse
   Appearance/Colour: Waxy/white/ivory
   Total Monoester: min. 40%
   HLB: 2,8
   Rise Melting Point Value: 44-46
   Remarks: Pharmacopee Francaise

TEGIN P 412:
   Propylene glycol monostearate, nse
   Appearance/Colour: Waxy/white/ivory
   Total Monoester: min. 50%
   HLB: 2,8
   Rise Melting Point Value: 33-36
   Remarks: Pharmacopee Francaise

GOLDSCHMIDT CHEMICAL CORP.: Products for Pharmaceutical
   Preparations: Solubilizers/Co-Emulsifiers:

Product:

TAGAT S2:
   Polyoxyethylene glycerol monostearate
   Appearance/Colour: solid, partially liquid/ivory
   Moles Ethylene Oxide: 20
   HLB: 15,0
   Hydroxyl Value: 65-85
   Remarks: Deutscher Arzneimittel-Codex 1979
            1. Erg 81

TAGAT R40:
   Ethylene oxide derivative of hydrogenated castor oil
   Appearance/Colour: solid/ivory
   Moles Ethylene Oxide: 40
   HLB: 15,0
   Hydroxyl Value: 55-80
   Remarks: Deutscher Arzneimittel-Codex 1979
            1. Erg 81

TAGAT O2:
   Polyoxyethylene glycerol monooleate
   Appearance/Colour: liquid/yellow
   Moles Ethylene Oxide: 20
   HLB: 15,0
   Hydroxyl Value: 70-85
   Remarks: Deutscher Arzneimittel-Codex 1979
            3. Erg. 83

TAGAT L2:
   Polyoxyethylene glycerol monolaurate
   Appearance/Colour: liquid/ivory
   Moles Ethylene Oxide: 20
   HLB: 15,7
   Hydroxyl Value: 60-80
   Remarks: Deutscher Arzneimittel-Codex 1979
            3. Erg. 83

GRINDSTED PRODUCTS, INC.: Emulsifiers:

AMIDAN:

ES   (powder*)

ARTODAN:

SP55   (powder*)

CETODAN:

50-00P   (block*)
50-CB   (block*)
70-00P   (block*)
90-40   (liquid)
95-CO   (liquid)

DIMODAN (animal):

PM   (beads)
S   (block)

DIMODAN (vegetable):

PV   (beads*)
PV-FF   (beads*)
PV300   (powder*)
PVP   (beads*)
CP   (plastic*)
LS   (soft plastic*)
O   (block*)
BP   (block*)

EMULDAN:

HAB40   (beads)
HAB52   (beads)
HV40   (beads*)
HV52   (beads*)
DG60   (beads*)
DG60   (liquid*)

* Kosher products available

GRINDSTED PRODUCTS, INC.: Emulsifiers(Continued):

FAMODAN:

MS (powder*)
TS (powder*)

LACTODAN:

P22 (beads*)

LIPODAN:

DE80 (beads*)
LM40 (beads*)
MS2   (beads*)
PS   (beads*)
STO   (beads*)

PANODAN:

FDP (powder*)
SD   (semi-liquid*)
150 (powder*)
125 (powder*)

PROMODAN:

SPV (beads*)
70   (beads*)

TRIODAN:

55 (powder)
20 (viscous liquid)

* Kosher products available

GUMIX INTERNATIONAL: Emulsifying Agents:

Agar Agar:

Introduction and Source:
Agar is a dried hydrophilic colloidal polygalactoside extracted from certain marine algae of the class Rhodophyceae. Commercially, two types of agar are produced from the Gelidium and Gracilaria species of red algae.

Physical Characteristics:
Agar is commercially available in bundles of thin, membraneous, agglutinated strips, usually less than 1 centimeter in width and 30-40 centimeters in length, or in cut, flaked, granulated, or powdered forms. It is white to pale yellow in color, is either odorless or has a slight characteristic odor, and has a mucilaginous taste.

Chemical Characteristics:
Chemically, agar is a salt of the sulfuric acid ester of a linear galactose polymer. Calcium, magnesium, sodium, and potassium seem to be basic constituents of the complex.

Applications:
The many varied applications of agar are outgrowths of its basic characteristics as a gelling agent, suspending agent, emulsifier, and bulking agent.

Guar Gum:

Introduction and Source:
Guar gum is derived from the ground endosperm of the guar plant, Cyamopsis tetragonolobus, of the Leguminosae family. The plant is cultivated commercially in India and Pakistan for human and animal consumption.

Physical Characteristics:
Guar gum is a white to yellowish-white, nearly odorless powder with a bland taste. The technical grades are slightly darker in color. Mesh sizes are readily available from 100 to 250.

Chemical Characteristics:
Guar gum, like locust bean gum, is a polysaccharide having a straight chain of D-mannopyranose units joined by β-(1-->4) linkages with a side-branching unit of single D-galactopyranose unit joined to every other mannose unit by α-(1-->6) linkages. The molecular weight of this galactomannan is 220,000.

Uses:
Guar gum uses are based primarily on thickening aqueous solutions and controlling the mobility of dispersed or solubilized materials in the water as well as water itself.

GUMIX INTERNATIONAL: Gum Arabic:

Introduction and Source:
Gum Arabic or acacia is the dried, gummy exudate from the stems or branches of Acacia senagal or of related species of Acacia. The trees which produce commercial grades of gum arabic grow primarily in the Sudan and Senagal regions of Africa.

Physical Characteristics:
Gum arabic tears, crystals, granules and powders are almost odorless and tasteless. The color ranges from white to yellowish white. The lighter the color, the better the quality of gum.

Chemical Characteristics:
Gum arabic is described as a complex calcium, magnesium, and potassium salt of arabic acid. The molecular weight of gum arabic is of the order of 250,000. It has a main backbone chain of (1-->3)-linked D-galactopyranose units, some of which are substituted at the C-6 position with various side-chains.

Uses:
The uses of gum arabic depend upon its action as a protective colloid or stabilizer and the adhesiveness of its water solutions. Its major use is in the food industry to impart viscosity, body, and texture to a variety of foods. It is nontoxic, odorless, colorless, tasteless, and completely water-soluble. It does not affect the odor, flavor, or color of foods.

Types Available:
    Spray Dried Gum Arabic Type A-180
    Spray Dried Gum Arabic Type A-230
    Powdered Gum Arabic Type B-100
    Powdered Gum Arabic Type B-200
    Powdered Gum Arabic Type E-920
    Granular Gum Arabic Type A-1
    Granular Gum Arabic Type A-2
    Granular Gum Arabic Type A-3
    Gum Arabic Type M-60
    Clean Amber Sorts
    Hand Picked Selected

GUMIX INTERNATIONAL: Gum Tragacanth:

Introduction and Source:
Gum tragacanth is the dried gum exuded by the stem elements of Astragalus gummifer or other Asiatic species of Astragalus. This plant is a small, low, bushy perennial shrub characterized by a relatively large tap root, which, along with the branches, is tapped for the gum.

Physical Characteristics:
Powdered ribbon gum tragacanth is white to light yellow in color, while powdered flake gum is yellowish white to tan. The gum is odorless and has an insipid taste.

Chemical Characteristics:
Gum tragacanth consists of a mixture of water-insoluble polysaccharides, bassorin which constitutes about 60-70% of the gum, and water-soluble polysaccharides, tragacanthin, which yield on hydrolysis, L-arabinose, L-fucose, D-xylose, D-galactose, and D-galacturonic acid. These acidic components are largely present as calcium, magnesium, and potassium salts.

Uses:
The uses of gum tragacanth depend on its action as an effective emulsifying and suspending agent with an extremely long shelf life and stability to heat and acidity.
    Food
    Pharmaceuticals
    Cosmetics
    Industrial

Types Available:
    Gum Tragacanth Ribbons and Flakes--all grades
    Powdered Gum Tragacanth Type G-1 NF
    Powdered Gum Tragacanth Type G-2 NF
    Powdered Gum Tragacanth Type G-2S NF
    Powdered Gum Tragacanth Type B-12 NF
    Powdered Gum Tragacanth Type B-1 NF
    Powdered Gum Tragacanth Type M-3 NF
    Powdered Gum Tragacanth Type C-5 NF
    Powdered Gum Tragacanth BP

GUMIX INTERNATIONAL: Locust Bean Gum:

Introduction and Source:
  Locust bean or carob gum is processed from the seeds of the locust bean tree known as Ceratonia siliqua which is widely cultivated in the Mediterranean area. The seeds are in a pod 10-20 cm long. These seeds or kernels, are the commercial source of locust bean gum, though only part of the seed is useful for that purpose.

Physical Characteristics:
  Locust bean gum is off-white to very light tan. The powder is processed in various sizes from about 50 mesh to 200 mesh. The highest grades are in a form of a near-white powder free from specks.

Chemical Characteristics:
  Locust bean gum is a polysaccharide built up of a main chain of mannose units with short branches of single galactose units. The molecular weight of locust bean gum is 310,000.

Uses:
  Locust bean gum is extremely versatile as a thickener or viscosity modifier, binder of free water, suspending agent and stabilizer.
    Food
    Industrial
    Pharmaceuticals and Cosmetics

Types Available:
    Locust Bean Gum Type A-100
    Locust Bean Gum Type A-250
    Locust Bean Gum Type A-270

HARCROS CHEMICALS INC.: Emulsifiers:

Product Name:

T-MULZ A02:
    Ionogenic Class: Nonionic Blend
    Physical Appearance: Light Amber Liquid
    Pour Point: 20F
    Typical Applications: Crop oil emulsifiers
    Federal Regulation Compliance: 180.1001d

T-MULZ COC:
    Ionogenic Class: Nonionic Blend
    Physical Appearance: Light Amber Liquid
    Pour Point: <10F
    Typical Applications: Crop oil concentrates
    Federal Regulation Compliance: 180.1001 c,d

T-MULZ FCO:
    Ionogenic Class: Nonionic Blend
    Physical Appearance: Light Amber Liquid
    Pour Point: <0F
    Typical Applications: Florida citrus oils
    Federal Regulation Compliance: 180.1001 c,d,e

T-MULZ FGO-17A:
    Ionogenic Class: Nonionic Blend
    Physical Appearance: Dark Amber Liquid
    Pour Point: 20F
    Typical Applications: Crop oil concentrates
    Federal Regulation Compliance: 180.1001 c,d

T-MULZ Mal5:
    Ionogenic Class: Anionic/Nonionic Blend
    Physical Appearance: Dark Amber Liquid
    Pour Point: 25F
    Typical Applications: Malathion formulations 5#/gallon
    Federal Regulation Compliance: 180.1001 c,d,e

T-MULZ PB:
    Ionogenic Class: Nonionic Blend
    Physical Appearance: Light Amber Liquid
    Pour Point: <0F
    Typical Applications: Pyrethrin formulations
    Federal Regulation Compliance: 180.1001 c,d,e

T-MULZ O:
    Ionogenic Class: Anionic/Nonionic Blend
    Physical Appearance: Dark Amber Liquid
    Pour Point: 55F
    Typical Applications: Matched pair with T-MULZ W for pesticides
    Federal Regulation Compliance: 180.1001 c,d,e

HARCROS CHEMICALS INC.: Emulsifiers(Continued):

Product Name:

T-MULZ W:
   Ionogenic Class: Anionic/Nonionic Blend
   Physical Appearance: Amber Liquid
   Pour Point: 64F
   Typical Applications: Matched pair with T-MULZ O for pesticides
   Federal Regulation Compliance: 180.1001 c,d,e

T-MULZ VO:
   Ionogenic Class: Nonionic Blend
   Physical Appearance: Light Amber Liquid
   Pour Point: <0F
   Typical Applications: Vegetable crop oil emulsifier
   Federal Regulation Compliance: 180.1001 c,d,e

T-MULZ 63:
   Ionogenic Class: Anionic/Nonionic Blend
   Physical Appearance: Dark Amber Liquid
   Pour Point: 30F
   Typical Applications: Parathion formulations
   Federal Regulation Compliance: 180.1001 c,d,e

T-MULZ 339:
   Ionogenic Class: Nonionic Blend
   Physical Appearance: Light Amber Liquid
   Pour Point: 23F
   Typical Applications: Dimethoate formulations
   Federal Regulation Compliance: 180.1001 c,d,e

T-MULZ 391:
   Ionogenic Class: Nonionic Blend
   Physical Appearance: Light Amber Liquid
   Pour Point: <0F
   Typical Applications: Oil based formulations
   Federal Regulation Compliance: 180.1001 c,d,e

T-MULZ 392:
   Ionogenic Class: Nonionic Blend
   Physical Appearance: Light Amber Liquid
   Pour Point: <0F
   Typical Applications: Oil based formulations
   Federal Regulation Compliance: 180.1001 c,d,e

T-MULZ 808A:
   Ionogenic Class: Nonionic Blend
   Physical Appearance: Light Amber Liquid
   Pour Point: <10F
   Typical Applications: Crop oil concentrates
   Federal Regulation Compliance: 180.1001 c,d,e

HARCROS CHEMICALS INC.: Emulsifiers(Continued):

Product Name:

T-MULZ 979:
   Ionogenic Class: Anionic/Nonionic Blend
   Physical Appearance: Dark Amber Liquid
   Pour Point: 70F
   Typical Applications: Single blend general purpose emulsifier for pesticide formulations.
   Federal Regulation Compliance: 180.1001 c,d,e

T-MULZ 8015:
   Ionogenic Class: Nonionic Blend
   Physical Appearance: Clear Liquid
   Pour Point: 0F
   Typical Applications: Spreader sticker
   Federal Regulation Compliance: 180.1001 c,d,e

T-MULZ AS-1151:
   Ionogenic Class: Anionic/Nonionic Blend
   Physical Appearance: Dark Amber Liquid
   Pour Point: 25F
   Typical Applications: Phenoxy herbicide formulations
   Federal Regulation Compliance: 180.1001 c,d,e

T-MULZ AS-1152:
   Ionogenic Class: Anionic/Nonionic Blend
   Physical Appearance: Dark Amber Liquid
   Pour Point: 35F
   Typical Applications: Phenoxy herbicide formulations
   Federal Regulation Compliance: 180.1001 c,d,e

T-MULZ AS-1153:
   Ionogenic Class: Anionic/Nonionic Blend
   Physical Appearance: Dark Amber Liquid
   Pour Point: 40F
   Typical Applications: Phenoxy herbicide formulations
   Federal Regulation Compliance: 180.1001 c,d,e

Crystal Inhibitor #5:
   Ionogenic Class: Nonionic Blend
   Physical Appearance: Amber Liquid
   Pour Point: 60F
   Typical Applications: Retards formation of crystals in 2,4-D Amine formulation dilutions
   Federal Regulation Compliance: 180.1001 c,d,e/175.105

CASUL 70HF:
   Ionogenic Class: Anionic
   Physical Appearance: Dark Amber Liquid
   Pour Point: 46F
   Typical Applications: 70% active high flash calcium sulfonate for use in agricultural emulsifiers
   Federal Regulation Complaince: 180.1001 c,d,e

HARCROS CHEMICALS INC.: T-MULZ Phosphate Esters:

T-MULZ Phosphate Esters are anionic surfactants designed for exceptional effectiveness in a wide range of applications, including:
* Industrial Alkaline Cleaners
* Emulsion Polymerization
* Household Cleaners
* Agricultural Surfactants
* Dry Cleaning Formulations
* Oil Field Operations
* Textile Processes

Phosphate Esters:

T-MULZ 1158:
Ionogenic Class: Anionic Blend
Physical Appearance: Viscous Amber Liquid
Typical Pour Point: 50F
% Active (Minimum): 99.5
Acid Values Mg KOH/gm pH 5.5/pH 9.5: 85/157
Typical Applications: Mineral oil emulsifier, metal processing, agricultural, solvent cleaners, textile processing.
Federal Regulation Compliance: 180.1001 c,d,e

T-MULZ 426:
Ionogenic Class: Anionic Blend
Physical Appearance: Viscous Amber Liquid
Typical Pour Point: 25F
% Active (Minimum): 99.5
Acid Values Mg KOH/gm pH 5.5/pH 9.5: 97/127
Typical Applications: Solvent degreasers, textile lubricant, agricultural.
Federal Regulation Compliance: 180.1001 c,d,e/175.105

T-MULZ 598:
Ionogenic Class: Anionic Blend
Physical Appearance: Viscous Amber Liquid
Typical Pour Point: 23F
% Active (Minimum): 99.5
Acid Values Mg KOH/gm pH 5.5/pH 9.5: 65/118
Typical Applications: Dry cleaning, general purpose cleaners, agricultural formulations, down hole scale inhibitor, emulsifiers
Federal Regulation Compliance: 180.1001 c,d,e/175.105

T-MULZ 565:
Ionogenic Class: Anionic Blend
Physical Appearance: Viscous Amber Liquid
Typical Pour Point: 36F
% Active (Minimum): 99.5
Acid Values Mg KOH/gm pH 5.5/pH 9.5: 62/115
Typical Applications: Industrial cleaners, dry cleaning, agriculture, petroleum lubricant, emulsifiers
Federal Regulation Compliance: 180.1001 c,d,e/178.3400

HARCROS CHEMICALS INC.: T-MULZ Phosphate Esters(Continued):

Phosphate Esters:

T-MULZ 734-2:
    Ionogenic Class: Anionic Blend
    Physical Appearance: Viscous Amber Liquid
    Typical Pour Point: 50F
    % Active (Minimum): 99.5
    Acid Values Mg KOH/gm pH 5.5/pH 9.5: 107/289
    Typical Applications: Built liquid detergents, agricultural tank mix additive.
    Federal Regulation Compliance: 180.1001 c,d,e

T-MULZ 800:
    Ionogenic Class: Anionic Blend
    Physical Appearance: Amber Liquid
    Typical Pour Point: 40F
    % Active (Minimum): 99.5
    Acid Values Mg KOH/gm pH 5.5/pH 9.5: 116/203
    Typical Applications: Alkaline cleaners, textile, hydrotrope, concentrated electrolyte solutions

T-MULZ 844:
    Ionogenic Class: Anionic Blend
    Physical Appearance: Amber Liquid
    Typical Pour Point: 25F
    % Active (Minimum): 80.0
    Acid Values Mg KOH/gm pH 5.5/pH 9.5: 156/300
    Typical Applications: Alkaline cleaners, textile, hydrotrope

T-MULZ 596:
    Ionogenic Class: Anionic Blend
    Physical Appearance: Amber Solid
    Typical Pour Point: 110F
    % Active (Minimum): 97.0
    Acid Values Mg KOH/gm pH 5.5/pH 9.5: 103/206
    Typical Applications: Emulsion polymerization
    Federal Regulation Compliance: 175.105/178.3400

Application/Benefits:
    Hydrotrope
    Agricultural
    Textile
    Alkaline Cleaners
    Emulsion Polymerization
    Industrial Applications

HENKEL CORP.: Emulsifiers:

Product Name:

CUTINA E24:
    PEG-20-Glyceryl Stearate
    Applications: O/W emulsifier especially suited for baby and children's care protection products; sun preparations

DEHYMULS E:
    Sorbitan Sesquioleate (and) Pentaerythritol Cocoate (and) Stearyl Citrate (and) Beeswax (and) Aluminum Stearate
    Applications: W/O emulsifiers with a high degree of water absorbency, good resistance to temperature fluctuations, hydrocarbon-free ointments

DEHYMULS F:
    Microcrystalline Wax (and) Pentaerythritol Cocoate (and) Stearyl Citrate (and) Glyceryl Oleate (and) Aluminum Stearate (and) Propylene Glycol
    Applications: W/O emulsifier with a high degree of water absorbency, good resistance to temperature fluctuations for soft creams

EUMULGIN B1:
    Ceteareth-12

EUMULGIN B2:
    Ceteareth-20

EUMULGIN B3:
    Ceteareth-30

EUMULGIN L:
    PPG-2-Ceteareth-9

Applications: Series of non-ionic emulsifiers which can be blended to obtain special emulsification effects (O/W emulsions). Soluble in alcohols, hydrocarbons, and most organic solvents. Solubility or dispersibility in water increases with degree of ethoxylation. Stable in many alkalis and acids under extreme pH conditions. Apart from emulsification, these ethoxylates offer unique characteristics of emolliency, body and conditioning.

HENKEL CORP.: Emulsifiers(Continued):

Product Name:

EUMULGIN M8:
    Oleth-10 (and) Oleth-5

EUMULGIN 05:
    Oleth-5

EUMULGIN 010:
    Oleth-10

   Applications: Series of non-ionic emulsifiers which can be blended to obtain special emulsification effects (O/W emulsions). Soluble in alcohols, hydrocarbons, and most organic solvents. Solubility or dispersibility in water increases with degree of ethoxylation. Stable in many alkalis and acids under extreme pH conditions. Apart from emulsification, these ethoxylates offer unique characteristics of emolliency, body and conditioning.

GENEROL 122:
    Soya Sterol
    Applications: Emollient, auxiliary or primary emulsifier, emulsion stabilizer and viscosity modifier

GENEROL 122 E5:
    PEG-5 Soya Sterol
    Substantive to hair, emollient and emulsifier, modifier and emulsion stabilizer

GENEROL 122 E10:
    PEG-10 Soya Sterol
    Emollient, auxiliary emulsifier, substantive to hair, emulsion stabilizer and viscosity modifier

GENEROL 122 E16:
    PEG-16 Soya Sterol
    Emollient, solubilizer and pigment wetter

GENEROL 122 E25:
    PEG-25 Soya Sterol
    Solubilizer and emollient, pigment wetter and deflocculating agent

LANETTE E:
    Sodium Cetearyl Sulfate
    O/W emulsifier for creams and ointments, especially in combination with fatty alcohols such as LANETTE 14, 16 and 18.

HERCULES INC.: HERCULES AR150 and AR160 Surfactants:

Low-Foaming, Nonionic Surface-Active Agents

HERCULES AR150 and AR160 surfactants are nonionic surface-active agents derived by ethoxylation of pale wood rosin. The first two digits of the product designation indicate the number of moles of ethylene oxide adducted. These polyethylene glycol esters of rosin are stable in alkaline and acidic media and are unaffected by hard water or the presence of metal ions. Their low-foaming properties and ability to control foaming make them well suited for use in industrial applications where a non-foaming detergent, emulsifier, or wetting agent is needed.

Typical Properties:
AR150:
    Physical form at 24C: liquid to soft wax
    Color, Gardner, at 40C: 13
    Viscosity at 38C, cps: 300
    pH, 1% solution: 9
    Cloudpoint, 2% solution, C: 60
    Surface tension, 0.1% in distilled water, dynes/cm: 39

AR160:
    Physical form at 24C: soft wax
    Color, Gardner, at 40C: 14
    Viscosity at 38C, cps: 250
    pH, 1% solution: 8
    Cloudpoint, 2% solution, C: 68
    Surface tension, 0.1% in distilled water, dynes/cm: 37.6

Outstanding Properties:
    Effective at low concentrations; low-foaming; compatible with anionic and cationic systems; stable to acids and alkalies; retain efficiency in the presence of metal ions.

Typical Uses:
    HERCULES AR150 and AR160 surfactants are useful wherever detergent, emulsifying, suspending, and/or dispersing action is needed. They offer a broad range of properties for use in industrial cleaners and detergents, and they are also used in suspending or dispersing operations where no foaming is essential.

FDA Status:
    HERCULES surfactants AR150 and AR160 are in compliance with requirements of the U.S. Food and Drug Administration for use in materials contacting foods as specified in the Code of Federal Regulations, Title 21, subject to the limitations and requirements of each regulation under the following Sections:
    175.105   Adhesives
    176.200   Defoaming agents used in coatings
    176.210   Defoaming agents used in the manufacture of paper and paperboard
    177.1200   Cellophane

HUMKO CHEMICAL DIVISION: ATMOS and ATMUL Glycerol Esters:

Product:

ATMOS 150:
   Color & Form @ 72F: Ivory White----Beads
   Free Glycerol %: 1.5 max
   Alpha Monoglyceride Content %: 52 min
   Total Monoglyceride Content %: 62
   H.L.B.: 3.5

ATMOS 300:
ATMOS 300K:
   Color & Form @ 72F: Clear Yellow----Fluid
   Free Glycerol %: 1.5 max
   Alpha Monoglyceride Content %: 46 min
   Total Monoglyceride Content %: 55
   H.L.B.: 2.8

ATMUL 84:
ATMUL 84K:
   Color & Form @ 72F: Ivory White----Beads or Flakes
   Free Glycerol %: 1.0 max
   Alpha Monoglyceride Content %: 40 min
   Total Monoglyceride Content %: 48
   H.L.B.: 2.8

ATMUL 124:
ATMOS 150K:
   Color & Form @ 72F: Ivory White----Flakes
   Free Glycerol %: 1.5 max
   Alpha Monoglyceride Content %: 52 min
   Total Monoglyceride Content %: 62
   H.L.B.: 3.5

ATMUL 695:
ATMUL 695K:
   Color & Form @ 72F: Clear Yellow----Fluid
   Free Glycerol %: 1.5 max
   Alpha Monoglyceride Content %: 52 min
   Total Monoglyceride Content %: 62
   H.L.B.: 3.0

HUMKO CHEMICAL DIVISION: Food Emulsifiers:

Mono- and Diglyceride Emulsifiers:

Product:

ATMUL 80:
    Color & Form @ 72F: Ivory White----Plastic Solid
    Alpha Monoglyceride Content, %: 40
    Approx. Total Monoglyceride Content, %: 48
    H.L.B.: 2.8
    Approximate Melting Point F: 115

ATMUL 84/84K:
    Color & Form @ 72F: Ivory White----Beads or Flakes
    Alpha Monoglyceride Content, %: 40
    Approx. Total Monoglyceride Content, %: 48
    H.L.B.: 2.8
    Approximate Melting Point F: 140 ATMUL 84
                                       144 ATMUL 84K

ATMUL 86K:
    Color & Form @ 72F: Ivory White----Plastic Solid
    Alpha Monoglyceride Content, %: 40
    Approx. Total Monoglyceride Content, %: 48
    H.L.B.: 2.8
    Approximate Melting Point F: 117

ATMUL 122:
    Color & Form @ 72F: Ivory White----Plastic Solid
    Alpha Monoglyceride Content, %: 52
    Approx. Total Monoglyceride Content, %: 62
    H.L.B.: 3.5
    Approximate Melting Point F: 126

ATMUL 124:
    Color & Form @ 72F: Ivory White----Flakes
    Alpha Monoglyceride Content, %: 62
    Approx. Total Monoglyceride Content, %: 62
    H.L.B.: 3.5
    Approximate Melting Point F: 140

ATMUL 500:
    Color & Form @ 72F: Ivory White----Plastic Solid
    Alpha Monoglyceride Content, %: 52
    Approx. Total Monoglyceride Content, %: 62
    H.L.B.: 3.5
    Approximate Melting Point F: 132

HUMKO CHEMICAL DIVISION: Food Emulsifiers(Continued):

Product:

ATMUL 651K:
   Color & Form @ 72F: Ivory White----Plastic Solid
   Alpha Monoglyceride Content, %: 52
   Approx. Total Monoglyceride Content, %: 62
   H.L.B.: 3.5
   Approximate Melting Point F: 133

ATMOS 150:
   Color & Form @ 72F: Ivory White
   Alpha Monoglyceride Content, %: 52
   Approx. Total Monoglyceride Content, %: 62
   H.L.B.: 3.5
   Approximate Melting Point F: 140

ATMOS 150K:
ATMOS 918K:
   Color & Form @ 72F: Ivory----Flakes (ATMOS 150K)
   Color & Form @ 72F: Ivory White----Beads (ATMOS 918K)
   Alpha Monoglyceride Content, %: 52
   Approx. Total Monoglyceride Content, %: 62
   H.L.B.: 3.5
   Approximate Melting Point F: 149

ATMUL 423K:
   Color & Form @ 72F: Clear Golden----Fluid
   Alpha Monoglyceride Content, %: 52
   Approx. Total Monoglyceride Content, %: 62
   H.L.B.: 3.5

ATMUL 300/300K:
   Color & Form @ 72F: Clear Yellow Fluid (May cloud at low temperatures)
   Alpha Monoglyceride Content, %: 46
   Approx. Total Monoglyceride Content, %: 55
   H.L.B.: 3.2
   Approximate Melting Point F: Clear Point:
     76 ATMOS 300
     79 ATMOS 300K

HUMKO CHEMICAL DIVISION: Food Emulsifiers(Continued):

Synergistic Emulsifiers Blends:

Product:

ATMUL 600H:
   Color & Form @ 72F: Ivory White----Plastic Solid
   Hydrate: Yes
   Alpha Monoglyceride Content %: 19
   Total Monoglyceride Content %: 23
   H.L.B.: 5.7
   Polysorbate 60%: 4.0
   Sorbitan Monostearate % (K): 0

ATMOS 729/729K:
   Color & Form @ 72F: Ivory White----Plastic Solid
   Hydrate: Yes
   Alpha Monoglyceride Content %: 12
   Total Monoglyceride Content %: 14
   H.L.B.: 6.7
   Polysorbate 60%: 11.3
   Sorbitan Monostearate % (K): 11.3

ATMOS 758:
   Color & Form @ 72F: Ivory White----Plastic Solid
   Hydrate: Yes
   Alpha Monoglyceride Content %: 5
   Total Monoglyceride Content %: 6
   H.L.B.: 7.5
   Polysorbate 60%: 13.5
   Sorbitan Monostearate % (K): 22.5

ATMOS 1069:
   Color & Form @ 72F: Ivory White----Plastic Solid
   Hydrate: Yes
   Alpha Monoglyceride Content %: 19
   Total Monoglyceride Content %: 24
   H.L.B.: 5.5
   Polysorbate 60%: 8.0
   Sorbitan Monostearate % (K): Sodium Stearoyl-2-Lactylate 4.0

TANDEM 5K:
   Color & Form @ 72F: Ivory White----Plastic Solid
   Hydrate: No
   Alpha Monoglyceride Content %: 31
   Total Monoglyceride Content %: 37
   H.L.B.: 8.1
   Polysorbate 60%: 40.0
   Sorbitan Monostearate % (K): 0

HUMKO CHEMICAL DIVISION: Food Emulsifiers(Continued):

Synergist Emulsifier Blends:
Product:
TANDEM 8:
   Color & Form @72F: Ivory White----Plastic Solid
   Hydrate: No
   Alpha Monoglyceride Content %: 31
   Total Monoglyceride Content %: 37
   H.L.B.: 8.1
   Polysorbate 60%: 40.0
   Sorbitan Monostearate % (K): 0
TANDEM 9:
TANDEM 12K:
   Color & Form @ 72F: Ivory White----Flake
   Hydrate: No
   Alpha Monoglyceride Content %: 33
   Total Monoglyceride Content %: 39
   H.L.B.: 6.0
   Polysorbate 60%: 25.0
   Sorbitan Monostearate % (K): 0
TANDEM 11H:
     11HK:
   Color & Form @ 72F: Ivory White----Plastic Solid
   Hydrate: Yes
   Alpha Monoglyceride Content %: 19.5
   Total Monoglyceride Content %: 25
   H.L.B.: 6.3
   Polysorbate 60%: 12.5
   Sorbitan Monostearate % (K): 0
TANDEM 22H:
     22HK:
   Color & Form @ 72F: Ivory White----Plastic Solid
   Hydrate: Yes
   Alpha Monoglyceride Content %: 19.5
   Total Monoglyceride Content %: 25
   H.L.B.: 7.2
   Polysorbate 60%: 18.0
   Sorbitan Monostearate % (K): 0
TANDEM 552/552K:
   Color & Form @ 72F: Clear Golden----Fluid
   Hydrate: Yes
   Alpha Monoglyceride Content %: 28
   Total Monoglyceride Content %: 32
   H.L.B.: 8.1
   Polysorbate 60%: 36.0
   Sorbitan Monostearate % (K): 0
TANDEM 100K:
   Color & Form @ 72F: Ivory White----Beads
   Hydrate: No
   Alpha Monoglyceride Content %: 32
   Total Monoglyceride Content %: 36
   H.L.B.: 5.2
   Polysorbate 60%: 0 (20% Polysorbate 80)
   Sorbitan Monostearate % (K): 0

HUMKO CHEMICAL DIVISON: HYSTRENE Lauric and Myristic Acids:

HYSTRENE fatty acids are characterized by light colors and excellent oxidative and thermal stability. These properties make the fatty acids adaptable in a wide range of applications, including use in personal care products, waxes, textile auxiliaries, pharmaceuticals, food grade products, emulsifiers, and a host of chemical intermediates.

Product:

INDUSTRENE 365:
  CTFA Adopted Name: Mixture Caprylic/Capric Acid
  Acid Value: 355-369
  Sap Value: 355-374

INDUSTRENE 325:
  CTFA Adopted Name: Coconut Acid
  Acid Value: 265-277
  Sap Value: 265-278

INDUSTRENE 223:
  CTFA Adopted Name: Hydrogenated Coconut Acid
  Acid Value: 266-274
  Sap Value: 267-276

INDUSTRENE 328:
  CTFA Adopted Name: Coconut Acid
  Acid Value: 253-260
  Sap Value: 253-261

HYSTRENE 5012:
  CTFA Adopted Name: Hydrogenated Coconut Acid
  Acid Value: 250-266
  Sap Value: 250-266

HYSTRENE 9512:
  CTFA Adopted Name: Lauric Acid
  Acid Value: 276-281
  Sap Value: 276-282

HYSTRENE 9912:
  CTFA Adopted Name: Lauric Acid
  Acid Value: 277-281
  Sap Value: 278-281

HYSTRENE 9014:
  CTFA Adopted Name: Myristic Acid
  Acid Value: 238-243
  Sap Value: 238-246

HYSTRENE 9514:
  CTFA Adopted Name: Myristic Acid
  Acid Value: 243-246
  Sap Value: 243-247

HUMKO CHEMICAL DIVISION: INDUSTRENE Stearic and Palmitic Acids:

INDUSTRENE fatty acids cover a large number of applications: alkyd resins, rubber compounding, mineral-flotation adjuvants, foam depressants, caulking compounds, lubricants, buffing compounds, water repellents, paper manufacture, polishes, metallic soaps, abrasives, metal treating compounds, candles, crayons, emulsifiers, and many more.

Product:

HYSTRENE 9016:
   CTFA Adopted Name: Palmitic Acid
   Acid Value: 216-220
   Sap Value: 216-221
INDUSTRENE 4518:
   CTFA Adopted Name: Stearic Acid
   Acid Value: 204-211
   Sap Value: 204-212
INDUSTRENE 5016:
   CTFA Adopted Name: Stearic Acid
   Acid Value: 207-210
   Sap Value: 207-211
HYSTRENE 5016:
   CTFA Adopted Name: Stearic Acid
   Acid Value: 206-210
   Sap Value: 206-211
INDUSTRENE 4516:
   CTFA Adopted Name: Palmitic Acid
   Acid Value: 205-211
   Sap Value: 205-212
INDUSTRENE R:
   CTFA Adopted Name: Stearic Acid
   Acid Value: 193-213
   Sap Value: 193-214
INDUSTRENE B:
   CTFA Adopted Name: Stearic Acid
   Acid Value: 199-207
   Sap Value: 199-208
INDUSTRENE 7018:
   CTFA Adopted Name: Stearic Acid
   Acid Value: 200-207
   Sap Value: 200-208
HYSTRENE 7018:
   CTFA Adopted Name: Stearic Acid
   Acid Value: 200-206
   Sap Value: 200-207
INDUSTRENE 9018:
   CTFA Adopted Name: Stearic Acid
   Acid Value: 196-201
   Sap Value: 196-202
HYSTRENE 9718:
   CTFA Adopted Name: Stearic Acid
   Acid Value: 196-201
   Sap Value: 196-202

HUMKO CHEMICAL DIVISION: 1,3-Propylenediamines:

The 1,3-propylenediamines are derived from hydrogenation of the cyanoethylated fatty primary amines. These products have both a primary and secondary amine group and have greater cationic character than the primary amine alone. Typical applications include use as gasoline detergents, corrosion inhibitors in petroleum products, dispersing aids and cationic asphalt emulsifiers.

Product:

KEMAMINE D-190:
  Description: N-90% Behenyl-Arachidyl
  % Diamine: 88 Typ

KEMAMINE D-999:
  Description: N-Oleic-Linoleic
  % Diamine: 88 Typ

Quaternary Ammonium Chlorides:

Water-miscible or -dispersible. They are extensively used as antistats, textile-softening agents, dyeing aids, corrosion inhibitors and emulsifiers.

Product:

KEMAMINE Q-2802C:
  CTFA Adopted Name: Dibehenyl Dimonium Chloride
  % Active Min: 75

KEMAMINE Q-9702C:
  CTFA Adopted Name: Quaternium-18
  % Active Min: 75

KEMAMINE Q-9743CHGW:
  CTFA Adopted Name: Tallow Trimonium Chloride
  % Active Min: 65

KEMAMINE Q-9743C:
  CTFA Adopted Name: Tallow Trimonium Chloride
  % Active Min: 65

KEMAMINE BQ-9742C:
  CTFA Adopted Name: Tallow Alkonium Chloride
  % Active Min: 75

HUMKO CHEMICAL DIVISION: Sorbitan Esters:

Product:

SMO 80:
SMO 80K:
   Color & Form @ 72F: Amber----Liquid
   Acid Value: 8 max
   Sap Value: 145-160
   Hydroxyl Value: 193-210
   H.L.B.: 4.3
   Viscosity @ 25C: Approx 1000 cps
   Water %: 1.0 max.

SML 20:
   Color & Form @ 72F: Amber
   Acid Value: 7 max
   Sap Value: 158-170
   Hydroxyl Value: 330-358
   H.L.B.: 8.6
   Viscosity @ 25C: Approx 4250 cps
   Water %: 1.0 max.

STO 85:
   Color & Form @ 72F: Liquid
   Acid Value: 15 max
   Sap Value: 170-190
   Hydroxyl Value: 55-70
   H.L.B.: 1.8
   Viscosity @ 25C: Approx 250 cps
   Water %: 1.0 max

SMS 60:
SMS 60K:
   Color & Form @ 72F: Ivory----Beads
   Acid Value: 5-10
   Sap Value: 147-157
   Hydroxyl Value: 235-260
   H.L.B.: 4.7
   Viscosity @ 25C: Solid
   Water %: 1.5 max

ICI SPECIALTY CHEMICALS: Emulsification Technology:

Emulsifier systems have been tailor-made by ICI Specialty Chemicals to provide a means of converting conventional paint resins into water-based emulsions.

Recommended Emulsifier System:

Resin Type: Short Alkyd
Oil: Tall
Emulsifier System: IL-2397
       HLB: 15
Emulsifier System: TWEEN 60
       HLB: 14.9

Resin Type: Short Alkyd
Oil: Soya
Emulsifier System: IL-2391
       HLB: 12.5
Emulsifier System: IL-2392
       HLB: 12.5

Resin Type: Medium Alkyd
Oil: Pure Oxidizing
Emulsifier System: IL-2393
       HLB: 15
Emulsifier System: TWEEN 80
       HLB: 15

Resin Type: Long Alkyd
Oil: Soybean
Emulsifier System: IL-2395
       HLB: 15
Emulsifier System: ATLAS G-4809
       HLB: 14.9

Resin Type: Polyester
Emulsifier System: IL-2395
       HLB: 15
Emulsifier System: IL-2393
       HLB: 15

Resin Type: Polyester (100% Solids)
Emulsifier System: ATLAS G 4809
       HLB: 14.9

ICI SPECIALTY CHEMICALS: Emulsification Technology(Continued):

Recommended Emulsifier Systems(Continued):

Resin Type: Epoxy
Emulsifier System: IL-2397
            HLB: 15
Emulsifier System: TWEEN 80
            HLB: 15

Resin Type: Urethane
Oil: Linseed
Emulsifier System: IL-2392
            HLB: 12.5
Emulsifier System: IL-2396
            HLB: 12.5

Resin Type: Chlorinated Paraffin
Emulsifier System: ATLOX 3409F
            HLB: 13

Resin Type: Hydrocarbon
Emulsifier System: G-3401
            HLB: 10

Resin Type: Drying Oils
Oil: 100%
Emulsifier System: G-1086
            HLB: 11.4

ICI SPECIALTY CHEMICALS: HYPERMER Polymeric Surfactants:

HYPERMER Surfactant:

A-60:
HLB: 5-7
% Active: 95
Primary Uses: w/o Emulsifier

A-95:
Viscosity poise @ 25C: 8-12
HLB: 6
% Active: 95
Primary Uses: o/w Emulsifier

A-109:
Viscosity poise @ 25C: 10-20
HLB: 13-15
% Active: 95
Primary Uses: o/w Emulsifier, Dispersant

A-394:
Viscosity poise @ 25C: 8-18
HLB: 7-9
% Active: 75
Primary Uses: o/w Emulsifier, Stabilizer, Dispersant

A-409:
Viscosity poise @ 25C: 6-12
HLB: 9-11
% Active: 97
Primary Uses: o/w Emulsifier, Stabilizer

B-246:
HLB: 5-6
% Active: 95
Primary Uses: w/o Emulsifier, Stabilizer

B-261:
HLB: 7-9
% Active: 95
Primary Uses: w/o, o/w Emulsifier, Stabilizer Dispersant

CG-6:
Viscosity poise @ 25C: 12-20
HLB: 11-12
% Active: 33
Primary Uses: Aqueous Dispersant, Emulsifier

ICI SURFACTANTS, INC.: HYPERMER Surfactants(Continued):

HYPERMER:

D-477:
   Viscosity poise @ 25C: 0.2
   HLB: 4
   % Active: 50
   Primary Uses: Oil Soluble Dispersant

E-475:
   HLB: 5-6
   % Active: 86
   Primary Uses: w/o, o/w Emulsifier, Stabilizer

E-476:
   HLB: 6-7
   % Active: 86
   Primary Uses: w/o, o/w Emulsifier, Stabilizer

E-488:
   HLB: 5-7
   % Active: 56
   Primary Uses: w/o, o/w Emulsifier Stabilizer

PRODUCT RANGE

HYPERMER A-60 Polymeric Surfactant:

A high molecular weight polyester surfactant supplied at a minimum of 95% active content in xylene.
HYPERMER A-60 is a polymeric nonionic surfactant showing outstanding solubility in a wide range of low solvency hydrocarbons and other low polarity liquids.
HYPERMER A-60 will also produce oil-in-water emulsions in combination with other nonionic surfactants of higher HLB and/or surface active agents such as aliphatic amines with oils of low solvency.

HYPERMER A-95 Polymeric Surfactant:

A modified polyester polymeric surfactant.

HYPERMER A-95 polymeric surfactant is a general purpose oil-in-water emulsifier. It is effective in emulsifying aliphatic and aromatic hydrocarbons as well as paraffinic oils.

ICI SPECIALTY CHEMICALS: HYPERMER Surfactants(Continued):

HYPERMER A-109 Polymeric Surfactant:

A modified polyester surfactant supplied as a 96% solids solution in xylene.

HYPERMER A-109 is a versatile water-soluble surface active polymer with many potential applications. It is a general purpose emulsifier for oil-in-water emulsions of a wide range of relatively high polarity liquids.
HYPERMER A-109 is an effective dispersant for a range of organic and inorganic solids in water.

HYPERMER A-394 Polymeric Surfactant:

A high molecular weight modified polyester surfactant supplied as a 75% solids solution in a mixture of white spirit and xylene.

HYPERMER A-394 is a highly efficient oil-in-water emulsifier/emulsion stabilizer.
HYPERMER A-394 is an effective dispersant for a wide range of organic and inorganic solids in organic liquids.

HYPERMER A-409 Polymeric Surfactant:

A modified polyester surfactant supplied at a minimum of 97% solids in xylene.

HYPERMER A-409 is an efficient, wide spectrum oil-in-water emulsifier/emulsion stabilizer for hydrocarbon oils and solvents, and other water-immiscible liquids of medium polarity.

HYPERMER B-246 Polymeric Surfactant:

An ABA type block copolymer, nonionic surfactant supplied at a minimum of 95% active in xylene.

HYPERMER B-246 is a very versatile surface active polymer with an average molecular weight of around 5,000. It is both an efficient emulsifier for many organic oil phases and a dispersant for a wide range of solid particles in aqueous or organic liquids.

ICI SPECIALTY CHEMICALS: HYPERMER Surfactants(Continued):

HYPERMER B-261 Polymeric Surfactant:

An ABA type block copolymer, nonionic surfactant supplied at a minimum of 95% active in xylene.
HYPERMER B-261, like its lower HLB variant HYPERMER B-246 is a versatile surface active polymer which has many applications as an emulsifier and dispersant.

HYPERMER CG-6 Polymeric Surfactant:

A 33% solids solution in 1/1:water/propylene glycol of an acrylic graft copolymer with nonionic surface active properties.
HYPERMER CG-6 is suggested primarily as a broad spectrum dispersant for both organic and inorganic solids in water.

HYPERMER D-477 Polymeric Surfactant:

An oil soluble dispersant for barite and drilling cuttings in invert emulsion oil based drilling fluids.
HYPERMER D-477 has been formulated specifically as an oil soluble dispersant for inorganic solids in invert emulsion oil-based drilling fluids.

HYPERMER E-475:

An oil soluble emulsifier stabilizer in xylene for use in both water-in-oil and oil-in-water emulsions and dispersant for hydrocarbon systems.
HYPERMER E-475 is an oil soluble emulsifier/dispersant, formulated specifically to stabilize oil-in-water emulsions of low polarity, high viscosity oils.
HYPERMER E-475 is a good dispersant for a wide range of solid particulate materials.

HYPERMER E-476:

An oil soluble emulsifier/stabilizer in xylene for oil-in-water or water-in-oil emulsions and dispersant for the hydrocarbon systems.
HYPERMER E-476 is an oil soluble emulsifier/dispersant.

HYPERMER E-488 Polymeric Surfactant:

An oil soluble emulsifier/stabilizer in xylene.
HYPERMER E-488 has outstanding solubility in a wide range of low polarity liquids.

JORDAN CHEMICAL CO.: Phosphate Esters:

All the JORDAPHOS surfactants are in the free acid form and 100% active anionics. They can be converted to a salt.

Trade Name:

JORDAPHOS 151:
Emulsifier for chlorinated solvents and aerosol propellents. Detergent surfactant.
Hydrophobe: Aromatic
Physical Form: Clear liquid
Acid Value Inflections: 1st: 45-55
                        2nd: 90-100

JORDAPHOS 236:
Emulsifier for aliphatic solvents. Coupling agent.
Hydrophobe: Aliphatic
Physical Form: Clear liquid
Acid Value Inflections: 1st: 90-100
                        2nd: 145-155

JORDAPHOS DT:
Surfactant for hard surface cleaning and textile scouring.
Hydrophobe: Aliphatic
Physical Form: Clear liquid
Acid Value Inflections: 1st: 70-90
                        2nd: 130-150

JORDAPHOS FDEO:
Low foaming surfactant. Coupling agent.
Hydrophobe: Blend
Physical Form: Clear liquid
Acid Value Inflections: 1st: 95-115
                        2nd: 180-200

JORDAPHOS JA-60:
Detergent surfactant and coupling agent. Compatible with inorganic builders.
Hydrophobe: Aliphatic
Physical Form: Clear viscous liquid
Acid Value Inflections: 1st: 105-115
                        2nd: 215-225

JORDAPHOS JB-40:
Oil and water soluble. Lubricity additive, rust inhibitor and emulsifier.
Hydrophobe: Aliphatic
Physical Form: Opaque viscous liquid
Acid Value Inflections: 1st: 70-90
                        2nd: 140-150

JORDAN CHEMICAL CO.: Phosphate Esters(Continued):

Trade Name:

JORDAPHOS JE-41:
 Emulsifiers for aromatic solvents. Detergents and solubilizer for hydrophilic surfactants in solvents.
   Hydrophobe: Aromatic
   Physical Form: Clear Viscous Liquid
   Acid Value Inflections: 1st: 85-100
                           2nd: 140-160

JORDAPHOS JE-51:
 Detergent material for emulsion polymerization.
   Hydrophobe: Aromatic
   Physical Form: Clear liquid
   Acid Value Inflections: 1st: 49-59
                           2nd: 85-100

JORDAPHOS JE-61:
 Detergent material for emulsion polymerization.
   Hydrophobe: Anionic
   Physical Form: Clear liquid
   Acid Value Inflections: 1st: 62-72
                           2nd: 110-125

JORDAPHOS JM-51:
 Emulsifier for aliphatic solvents, chlorinated solvents and aerosol propellents. Cutting oils and lubricants.
   Hydrophobe: Aromatic
   Physical Form: Hazy viscous liquid
   Acid Value Inflections: 1st: 45-55
                           2nd: 80-90

JORDAPHOS JM-71:
 Similar to JM-51 but water soluble. Emulsification of cutting oils.
   Hydrophobe: Aromatic
   Physical Form: Hazy viscous liquid
   Acid Value Inflections: 1st: 33-41
                           2nd: 60-80

JORDAPHOS JP-70:
 Water soluble lubricant. Non-foaming.
   Hydrophobe: Aromatic
   Physical Form: Clear liquid
   Acid Value Inflections: 1st: 85-90
                           2nd: 135-150

JORDAN CHEMICAL CO.: Phosphate Esters (Continued):

Trade Name:

JORDAPHOS JS-41:
  Textile wetting and alkaline scour. Salts used for dry cleaning.
  Hydrophobe: Aliphatic
  Physical Form: Hazy viscous liquid
  Acid Value Inflections: 1st: 95-115
                          2nd: 160-180

JORDAPHOS JS-61:
  Adds lubricity and load-bearing properties of water and oil based lubricants. Cutting oil emulsifier.
  Hydrophobe: Aliphatic
  Physical Form: Hazy viscous liquid
  Acid Value Inflections: 1st: 75-85
                          2nd: 125-150

JORDAPHOS JS-71:
  Alkali-stable detergent emulsifier, wetting agent dispersant.
  Hydrophobe: Aliphatic
  Physical Form: Opaque viscous liquid.
  Acid Value Inflections: 1st: 58-70
                          2nd: 100-120

JORDAPHOS RA-60:
  Emulsifier for aliphatic solvents. Coupling agent.
  Hydrophobe: Aliphatic
  Physical Form: Clear liquid
  Acid Value Inflections: 1st: 114-116
                          2nd: 180-185

LANAETEX PRODUCTS, INC.: Emulsifiers:

Product:

ANATOL:
Lanolin Alcohols
Color: Pale yellow
Sulfated Ash: 0.15 max.
Volatiles (1 hr. @ 105C): 0.5% max.
Acid Value: 2.0 max.
Saponification Value: 12.0 max.
Melting Point: 58C min.
CTFA: Lanolin Alcohol
Attributes: ANATOL is used as a W/O emulsifier for a wide variety of cosmetic products, giving them an emolliency to the skin and hair.

COSMETIC LANOLIN-ANHYDROUS USP:
Color: Extra pale yellow
Alkalinity: None
Ammonia: None
Water Soluble Oxidizable Substances: None
Petrolatum: None
Chlorides: USP
Water Soluble Acids or Alkalies: USP
Loss on Drying: 0.25% max.
Residue on Ignition: 0.1% max.
Acid Value: Max. 2cc N/10 NaOH (10 grams)
Free Fatty Acid (Oleic): 0.56% max.
Iodine Value (Hanus): 18-36
Melting Point: 36-42C
CTFA: Lanolin
Attributes: A naturally derived product possessing effective emollient action with emulsifying power. It is self emulsifying, forming a stable W/O emulsions and absorbs up to two or more times its weight in water. It is chemically nonreactive under normal conditions of use.

ETHLANA 12:
Ethoxylated Isostearyl Alcohol
Appearance: White to off white paste
Acid Value: 0.5 max.
Iodine Value (Hanus): 5.0 max.
pH (3% soln.): 6-8
Haze Point: (3% soln.): 70C
Hydroxyl Value: 70-80
CTFA: Isosteareth 12
Attributes: ETHLANA 12 can be used in hair preparations, bath oils, lotions, deodorants, or any other preparations where a nonionic emulsifier is needed.

LANAETEX PRODUCTS, INC.: Emulsifiers(Continued):

Product:

ETHLANA 22:
  Ethoxylated Isostearyl Alcohol
  Appearance: White to off white solid
  Acid Value: 0.5 max.
  Iodine value (Hanus): 5.0 max.
  pH (3% soln.): 6-8
  Haze Point (3% soln.): 60-70
  Hydroxyl Value: 100C
  CTFA: Isosteareth 22
  Attributes: ETHLANA 22 can be used in hair preparations, bath oils, lotions, deodorants, or any other preparations where a nonionic emulsifier is needed.

ETHLANA 50:
  Ethoxylated Isostearyl Alcohol
  Appearance: White to off white solid
  Acid Value: 0.5 max.
  Iodine Value (Hanus): 5.0 max.
  pH (3% soln.): 6-8
  Haze Point: 100C
  Hydroxyl Value: 35-45
  CTFA: Isosteareth 50
  Attributes: ETHLANA 50 is a product that can be used in hair preparations, bath oils, lotions, deodorants, or any other preparations where a nonionic emulsifier is needed.

ETHOXOL-3:
  Ethoxylated Oleyl Alcohol (3 moles)
  Appearance: Hazy liquid
  Odor: Mild, faint
  Water (%): 1.0 max.
  pH @ 25C (3% solution): 5-7
  Ash: 1.0 max.
  Acid Value: 2.0 max.
  Iodine value (Hanus): 57-62
  CTFA: Oleth 3
  Attributes: ETHOXOL-3 is an efficient solubilizer and clear gels at low ratios of emulsifier and oil, imparting emolliency, lubricity, and slip. Essentially the product can be used as an emulsifier, spreading agent, superfatting, and perfume solubilizer.

LANAETEX PRODUCTS, INC.: Emulsifiers(Continued):

Product:

ETHOXOL-10:
   Ethoxylated Oleyl Alcohol (10 moles)
   Appearance: Off-white, semi-solid
   Odor: Mild, faint
   Water (%): 1.0 max.
   pH @ 25C (3% solution): 5-7
   Ash: 0.5 max.
   Acid Value: 2.0 max.
   Iodine Value (Hanus): 31-37
   CTFA: Oleth 10
   Attributes: ETHOXOL-10 is an efficient solubilizer and clear gels at low ratios of emulsifier and oil, imparting emolliency, lubricity, and slip. Essentially the product can be used as an emulsifier, spreading agent, superfatting, and perfume solubilizer.

ETHOXOL-20:
   Ethoxylated Oleyl Alcohol (20 moles)
   Appearance: White-waxy solid
   Odor: Mild, faint
   Water (%): 3.0 max.
   pH @ 25C (3% solution): 5-7
   Ash: 2.0 max.
   Acid Value: 2.0 max.
   Iodine Value (Hanus): 18-25
   CTFA: Oleth 20
   Attributes: ETHOXOL-20 is an efficient solubilizer and clear gels at low ratios of emulsifier and oil, imparting emolliency, lubricity, and slip. Essentially the product can be used as an emulsifier, spreading agent, superfatting, and perfume solubilizer.

ETHOXOL-44:
   Ethoxylated Oleyl Alcohol (44 moles)
   Appearance: White - waxy solid
   Odor: Bland
   Water (%): 3.0 max.
   pH @ 25C (3% solution): 5-7
   Ash: 2.0 max.
   Acid Value: 2.0 max.
   Iodine Value (Hanus): 4-10
   CTFA: Oleth 44
   Attributes: ETHOXOL-44 is an efficient solubilizer and clear gels at low ratios of emulsifier and oil, imparting emolliency, lubricity, and slip. Essentially the product can be used as an emulsifier, spreading agent, superfatting, and perfume solubilizer.

LANAETEX PRODUCTS, INC.: Emulsifiers(Continued):

ETHOXOL-44-M:
Propylene Glycol and Ethoxylated Oleyl Alcohol
Appearance: Clear liquid (@ room temp.)
Odor: Bland
Acid Value: 2.0 max.
Iodine Value (Hanus): 2-8
pH @ 25C (3% soln): 5-7
CTFA: Propylene Glycol (and) Oleth-44
Attributes: ETHOXOL-44-M is an efficient solubilizer and clear gels at low ratios of emulsifier and oil, imparting emolliency, lubricity, and slip. Essentially the product can be used as an emulsifier, spreading agent, superfatting, and perfume solubilizer.

ETHOXYCHOL-24:
Polyethoxylated Cholesterol
Appearance: Light waxy solid
Odor: Mild, faint
Ash: 0.2% max.
Acid Value: 2.0 max.
pH @ 25C (10% aqueous soln.): 4.0-7.0
Saponification Value: 8.0 max.
CTFA: Choleth-24
Attributes: ETHOXYCHOL-24 being a non-ionic solubilizer, can be used as an emulsifier and wetting agent. It has the property of forming gels in vegetable oils.

EXTAN-GO:
Propylene Glycol and Glycerol Oleate
Appearance: Clear oily liquid, may cloud at low temperature.
Color: Yellow
Odor: Faint, pleasant
Acid Value: 3.0 max.
Specific Gravity @ 25C: 1.02 max.
Solubilities: Soluble in ethanol, isopropanol, cottonseed oil, and mineral oil. Insoluble in water.
CTFA: Glyceryl Oleate (and) Propylene Glycol
Attributes: The non-ionic ester is used primarily as an emulsifier and thickener in cosmetic products. In combination with sorbitol solution, it is very effective in forming a water-in-oil emulsions, having a smooth non-greasy characteristic feel.

LANAETEX PRODUCTS, INC.: Emulsifiers(Continued):

EXTAN-LT:
   Sorbitan Monolaurate
   Appearance: Liquid
   Color: Amber
   Specific Gravity @ 25C: 1.049-1.055
   Acid Value: 7.0 max.
   Saponification value: 158-170
   Hydroxyl Value: 330-358
   Moisture (%): 1.0 max.
   Viscosity, cps, 25C: 4500
   CTFA: Sorbitan Laurate
   Attributes: It is a simple ester of sorbitol and fatty acid. As a result of their lipophilic nature, it is used as W/O emulsifier, stabilizer, softener, bodying agent, and opacifier in cosmetic and pharmaceutical creams and lotions.

EXTAN-OT:
   Sorbitan Monooleate
   Appearance: Liquid
   Color: Amber
   Acid Value: 6-8
   Saponification Value: 149-160
   Hydroxyl Value: 193-209
   Moisture (%): 0.5 max.
   Specific Gravity @ 25C: 0.995-1.015
   Viscosity, cps, 25C: 1000
   CTFA: Sorbitan Oleate
   Attributes: A simple ester of sorbitol and oleic acid. Used as W/O emulsifier, stabilizer, softener, bodying agent, and opacifier in cosmetic and pharmaceutical creams and lotions.

EXTAN-PT:
   Sorbitan Monopalmitate
   Appearance: Free-flowing beads
   Color: Tan
   Acid Value: 4.0-7.5
   Saponification Value: 140-150
   Hydroxyl Value: 275-305
   Moisture (%): 1.5 max.
   CTFA: Sorbitan Palmitate
   Attributes: A simple ester of sorbitol and palmitic acid. It is used as W/O emulsifier, stabilizer, softener, bodying agent, and opacifier in cosmetic and pharmaceutical creams and lotions.

LANAETEX PRODUCTS, INC.: Emulsifiers(Continued):

EXTAN-SOT:
　　Sorbitan Sesquioleate
　　Appearance: Liquid
　　Color: Clear to yellow
　　Acid Value: 10 max.
　　Saponification Value: 150-160
　　Hydroxyl Value: 187-210
　　Moisture (%): 1.0 max.
　　CTFA: Sorbitan Sesquioleate
　　Attributes: A simple ester of sorbitol and oleic acid. Used as W/O emulsifier, stabilizer, softener, bodying agent, and opacifier in cosmetic and pharmaceutical creams and lotions.

EXTAN-ST:
　　Sorbitan Monostearate
　　Appearance: Free-flowing beads
　　Color: Cream
　　Acid Value: 5-10
　　Saponification Value: 147-157
　　Hydroxyl Value: 235-260
　　Moisture (%): 1.5 max.
　　CTFA: Sorbitan Stearate
　　Attributes: A simple ester of sorbitol and stearic acid. Used as W/O emulsifier, stabilizer, softener, bodying agent, and opacifier in cosmetic and pharmaceutical creams and lotions.

LANAETEX-A16:
　　Polyethylene (16) Lanolin Ether
　　Color: Amber
　　Odor: Faint
　　Acid Value: 1.0 max.
　　Saponification Value: 2-6
　　Hydroxyl Value: 60-75
　　Iodine Value (Hanus): 18-30
　　Moisture % (24 hrs. @ 105C): 1.0 max.
　　CTFA: Laneth-16
　　Attributes: LANAETEX-A16 is an excellent O/W emulsifier and an auxiliary W/O emulsifier. LANAETEX-A16 is an emollient solubilizer for perfumes and mineral and vegetable oils used in a wide range of cosmetic products.

LANAETEX PRODUCTS, INC.: Emulsifiers(Continued):

LANAETEX CLC:
  Emollient Absorption Base
  Appearance: Soft solid
  Color: Pale to yellow
  Odor: Faint, sterol-like
  Water Soluble Acids & Alkalies: None
  Moisture (%): 0.3 max.
  Acid Value: 1.0 max.
  Saponification Value: 1.0 max.
  Melting Point: 43-45C
  CTFA: Petrolatum, Lanolin (and) Lanolin Alcohol
  Attributes: An attractive application is in W/O emulsions.
LANAETEX CLC is extensively used as an auxiliary emulsifier, to enhance and toughen the interfacial film, thus providing an effective penetration. Small quantities will minimize irritation in alkaline emulsions.

LANAETEX-H:
  Absorption Base
  Color (Gardner): 8 max.
  Odor: Mild, sterol-like
  Appearance: Semi-solid
  Acid Value: 2.0 max.
  Melting Point: 50-56C
  Water Absorption, (%): 1500-2200
  CTFA: Petrolatum (and) Lanolin (and) Lanolin Alcohol (and)
        Sorbitan Sesquioleate (and) Beeswax
  Attributes: An attractive application is in W/O emulsions.
LANAETEX-H is extensively used as an auxiliary emulsifier, to enhance and toughen the interfacial film, thus providing an effective penetration. Small quantities will minimize irritation in alkaline emulsions.

LANAETEX L-15:
  Emollient Absorption Base
  Appearance: Soft solid
  Color: Yellow
  Odor: Sterol-like, mild
  Water Soluble Acids & Alkalies: None
  Moisture (%): 0.2 max.
  Acid Value: 2.0 max.
  Saponification Value: 18 max.
  Melting Point: 50-60C
  CTFA: Petrolatum (and) Lanolin (and) Lanolin Alcohol
  Attributes: An attractive application is in W/O emulsions.
LANAETEX L-15 is extensively used as an auxiliary emulsifier, to enhance and toughen the interfacial film, thus providing an effective penetration. Small quantities will minimize irritation in alkaline emulsions.

LANAETEX PRODUCTS, INC.: Emulsifiers(Continued):

LANOBASE SE:
  A Non-ionic Emulisifer
  Appearance: Soft solid
  Color: Slightly yellow, creamy, transparent when melted
  Odor: Practically none
  Acid Value: 3.0 max.
  Iodine Value (Hanus): 6 max.
  Saponification Value: 10 max.
  Melting Point: 35-41C
  Solubility: Soluble in hot alcohol. Soluble in hot oils (sediment on cooling). Insoluble in water - forms emulsions with water in all proportions.
  CTFA: PEG 6-32 (and) PEG-75 Lanolin
  Attributes: LANOBASE SE is a primary or auxiliary non-ionic emulsifier, solubilizer, plasticizer, and dispersing agent. It leaves a soft-nontacky after feel on the skin.

LANOLA 90:
  Lanolin-Glyco-Stearate-Lanolate
  Appearance: Semi-solid, lanolin-like
  Color: Light to medium amber
  Odor: Bland, lanolin-like, much milder
  Moisture (%): 0.25 max.
  Free Fatty Acids: 0.6% max.
  Melting Point: 34-42C
  Iodine Value (Hanus): 25-45
  Saponification Value: 90-120
  CTFA: Lanolin (and) PEG-8 Stearate
  Attributes: A non-ionic ester is compatible with anionic and cationic surfactants and functions as an emulsifier, plasticizer, lubricant, wetting, dispersing, and thickening agents. These properties are used in shampoos, cream rinses, and hand lotions.

LANOLIN ANHYDROUS USP:
  Color: Yellow
  Alkalinity: None
  Ammonia: None
  Water Soluble Oxidizable Substances: None
  Petrolatum: None
  Chlorides: USP
  Water Soluble Acids or Alkalies: USP
  Loss on Drying: 0.25% max.
  Residue on Ignition: 0.1% max.
  Acid Value: Max. 2cc N/10NaOH (10 grams)
  Free Fatty Acid (Oleic): 0.56% max.
  Iodine Value (Hanus): 18-36
  Melting Point: 36-42C
  CTFA: Lanolin
  Attributes: A naturally derived product possessing effective emollient action with emulsifying power. It is self-emulsifying, forming stable W/O emulsions and absorbs up to two or more times its weight in water. It is chemically non-reactive.

LANAETEX PRODUCTS, INC.: Emulsifiers(Continued):

LANOXIDE-52:
  Polyoxyethylene (40) Stearate
  Appearance: Ivory, waxy solid
  Odor: Bland
  Acid Value: 1.0 max.
  Moisture (%): 3.0 max.
  Saponification Value: 25-35
  Hydroxyl Value: 27-40
  CTFA: PEG-40 Stearate
  Attributes: LANOXIDE-52 is an auxiliary emulsifier, thickener, and solubilizer used in a variety of cosmetic products, such as creams and lotions.

LANTOX 55:
  Water and Alcohol Soluble Lanolin
  Appearance: Viscous liquid, gels at lower temperature.
  Color: Light to medium amber
  Odor: Mild, pleasant, characteristic
  Solubility: Soluble in water, 95% alcohol and alcohol/water mixture
  Specific Gravity @ 25C: 1.05-1.09
  Iodine Value (Hanus): 8 max.
  Saponification Value: 14 max.
  Water (%): 48-52
  CTFA: PEG-75 Lanolin
  Attributes: An alcohol and water soluble non-ionic surfactant with an unusual and unique stability over a wide pH range. It is compatible with various ionic agents used in cosmetic and pharmaceutical formulations. Emolliency and dermal softening properties characterized by the presence of Lanolin, are retained for use in skin and hair products.

LANTOX 110:
  Anhydrous Water and Alcohol Soluble Lanolin
  Appearance: Hard waxy solid
  Color: Amber
  Odor: Mild, pleasant characteristic
  Acid Value: 2.0 max.
  Iodine Value (Hanus): 14 max.
  Saponification Value: 20 max.
  CTFA: PEG-75 Lanolin
  Attributes: An alcohol and water soluble non-ionic surfactant with an unusual and unique stability over a wide range. It is compatible with various ionic agents used in cosmetic and pharmaceutical formulations. Emolliency and dermal softening properties characterized by the presence of Lanolin, are retained for use in skin and hair products.

LANAETEX PRODUCTS, INC.: Emulsifiers(Continued):

LANYCOL-30:
   Polyoxyethylene (4) Lauryl Ether
   Appearance: Colorless liquid
   Specific Gravity @ 25C: 0.950 approximately
   Acid Value: 2.0 max.
   pH @ 25C (1% aqueous soln.): 5.0-7.5
   Hydroxyl Value: 145-165
   CTFA: Laureth-4
   Attributes: Since the product doesn't contain ester linkages, it is stable to many alkalies and acids beyond the pH range which ordinary emulsifiers would be expected to withstand.

LANYCOL-35:
   Polyoxyethylene (23) Lauryl Ether
   Appearance: White solid
   Moisture (%): 1.0 max.
   Free Fatty Acid: 1.0 max.
   pH @ 25C: 5.0-7.5
   Hydroxyl Value: 40-55
   CTFA: Laureth-23
   Attributes: Since the product doesn't contain ester linkages, it is stable to many alkalies and acids beyond the pH range which ordinary emulsifiers would be expected to withstand.

LANYCOL-58:
   Polyoxyethylene (20) Cetyl Ether
   Appearance: White solid
   Acid Value: 1.5 max.
   Moisture (%): 3.0 max.
   Hydroxyl Value: 45-60
   CTFA: Ceteth-20
   Attributes: Since the product doesn't contain ester linkages, it is stable to many alkalies and acids beyond the pH range which ordinary emulsifiers would be expected to withstand. LANYCOL-58 is also an excellent gelling agent.

LANYCOL-92:
   Polyoxyethylene (2) Oleyl Ether
   Appearance: Yellow liquid
   Acid Value: 1.0 max.
   Moisture (%): 1.0 max.
   Hydroxyl Value: 160-180
   CTFA: Oleth-2
   Attributes: The above product has no ester linkage. It is nonionic in character and extremely stable over a wide pH range. The lipophilic emulsifier is a good solubilizer. LANYCOL-92 is an excellent gelling agent.

LANAETEX PRODUCTS, INC.: Emulsifiers(Continued):

LANYCOL-98:
    Polyoxyethylene (20) Oleyl Ether
    Appearance: Cream solid
    Moisture (%): 1.0 max.
    Free Fatty Acid: 1.0% max.
    pH @ 25C (20% aqueous soln.): 5.5-7.0
    Iodine Value (Hanus): 23 max.
    Hydroxyl Value: 50-65
    CTFA: Oleth-20
    Attributes: Since the product doesn't contain ester linkages, it is stable to many alkalies and acids beyond the pH range which ordinary emulsifiers would be expected to withstand. LANYCOL-98 is also an excellent gelling agent.

LAXAN-ESL:
    Polyoxyethylene (20) Sorbitan Monolaurate
    Appearance: Yellow liquid
    Acid Value: 2.0 max.
    Moisture (%): 3.0 max.
    Saponification Value: 40-52
    Hydroxyl Value: 95-114
    CTFA: Polysorbate-20
    Attributes: With hydrophilic properties, LAXAN-ESL is an excellent O/W emulsifier and solubilizer for perfumes, flavors, and colors.

LAXAN-ESM:
    Polyoxyethylene (5) Sorbitan Monooleate
    Appearance: Oily liquid
    Color: Amber
    Specific Gravity @ 25C: Approx. 1
    Acid Value: 2.0 max.
    Saponification Value: 96-104
    Hydroxyl Value: 134-150
    Moisture (%): 3.0 max.
    CTFA: Polysorbate 81
    Attributes: With hydrophilic properties, LAXAN-ESM is an excellent O/W emulsifier and solubilizer for perfumes and colors.

LAXAN-ESO:
    Polyoxyethylene (20) Sorbitan Monooleate
    Appearance: Yellow liquid
    Acid Value: 2.0 max.
    Moisture (%): 3.0 max.
    Saponification Value: 45-55
    Hydroxyl Value: 65-80
    CTFA: Polysorbate-80
    Attributes: With hydrophilic properties, LAXAN-ESO is an excellent O/W emulsifier and solubilizer for perfumes, flavors, and colors.

LANAETEX PRODUCTS, INC.: Emulsifiers(Continued):

LAXAN-ESP:
   Polyoxyethylene (20) Sorbitan Monopalmitate
   Appearance: Yellow liquid
   Acid Value: 2.0 max.
   Moisture (%): 3.0 max.
   pH (3% solution) @ 25C: 6.0-7.5
   Saponification Value: 41-52
   Hydroxyl Value: 90-105
   CTFA: Polysorbate-40
   Attributes: With hydrophilic properties, LAXAN-ESP is an excellent O/W emulsifier and solubilizer for perfumes, flavors and colors.

LAXAN-ESR:
   Polyoxyethylene (20) Sorbitan Tristearate
   Appearance: Waxy solid
   Color: Pale amber to amber
   Acid Value: 2.0 max.
   Saponification Value: 88-98
   Hydroxyl Value: 44-60
   Moisture (%): 3.0 max.
   CTFA: Polysorbate 65
   Attributes: LAXAN ESR, being a surfactant in character, is very hydrophilic and is used as an emulsifier, for oil in water preparations, for dispersing or solubilizing oils as well as making anhydrous oil-based products water-soluble or washable.

LAXAN-ESS:
   Polyoxyethylene (20) Sorbitan Monostearate
   Appearance: Yellow, oily liquid
   Acid Value: 2.0 max.
   Moisture (%): 3.0 max.
   Saponification Value: 45-55
   Hydroxyl Value: 80-96
   CTFA: Polysorbate-60
   Attributes: With hydrophilic properties, LAXAN-ESS, is an excellent O/W emulsifier and solubilizer for perfumes, flavors, and colors.

LAXAN-EST:
   Polyoxyethylene (20) Sorbitan Trioleate
   Appearance: Tan solid
   Acid Value: 2.0 max.
   Moisture (%): 5.0 max.
   Hydroxyl Value: 39-52
   Saponification Value: 82-95
   CTFA: Polysorbate-85
   Attributes: With hydrophilic properties, LAXAN-EST is an excellent O/W emulsifier and solubilizer for perfumes, flavors, and colors.

LANAETEX PRODUCTS, INC.: Emulsifiers(Continued):

LINSOL ETO:
Ethoxylated Acetylated Lanolin Alcohol
Appearance: Light amber liquid
Odor: Faint, pleasant
Specific Gravity @ 25C: 1.035-1.055
Acid Value: 3.0 max.
Hydroxyl Value: 55-75
Saponification Value: 65-75
CTFA: Laneth-10 Acetate
Attributes: LINSOL ETO is an important primary non-ionic O/W emulsifier, solubilizer, and wetting agent. These useful properties can be utilized in aqueous preparations as well as in powder products.

NAETEX-EF:
Oleamidopropyl Ethyldimonium Ethosulfate
Appearance: Viscous, liquid
Total solids: 90-99%
CTFA: Oleamidopropyl Dimethylamine Ethonium Ethosulfate
Attributes: The above product can be used in hair conditioning formulations, as a cream and lotion for the skin.

NAETEX 118:
Oleyl Dimethylaminopropylamido Glycolate
Appearance: Clear light amber liquid
% Total solids (4 hrs. @ 105C): 15-20
CTFA: Oleamidopropyl Dimethylamine Glycolate
Attributes: The above product can be used in hair conditioning formulations, as a cream and lotion for the skin.

NAETEX 135:
Oleyl Dimethylaminopropylamido Glycolate
Appearance: Clear amber liquid
% Total Solids (4 hrs. @ 25C): 32-37
CTFA: Oleamidopropyl Dimethylamine Glycolate
Attributes: The above product can be used in hair conditioning formulations, as a cream and lotion for the skin.

NAETEX 170:
Oleyl Dimethylaminopropylamido Glycolate
Appearance: Moderately viscous amber liquid
% Total Solids (3 hrs. @ 105C): 67-72
CTFA: Oleamidopropyl Dimethylamine Glycolate
Attributes: The above product can be used in hair conditioning formulations, as a cream and lotion for the skin.

LIPO CHEMICALS INC.: Emulsifiers:

LIPOSORB L:
CTFA Adopted Name: Sorbitan Laurate

Description:
LIPOSORB L consists predominantly of the monoesters of lauric acid and hexitol anhydrides derived from sorbitol. It is a weak nonionic w/o emulsifier for neutral and mildly alkaline and acid pH systems. It is compatible with anionic, cationic and nonionic surfactants.

Applications:
LIPOSORB L is used in cosmetics, toiletries and topical pharmaceuticals primarily as an auxiliary emulsifier and stabilizer in combination with LIPOSORB L-20.

Typical Properties:
Appearance: Yellow/amber liquid
Odor: Bland, characteristic
Ionic Character: Nonionic
HLB: 8.6
Acid Value: 7 max.
Hydroxyl Value: 330-360
Saponification Value: 158-170
Moisture, %: 1 max.

LIPOSORB P:
CTFA Adopted Name: Sorbitan Palmitate

Description:
LIPOSORB P consists predominantly of the monoesters of palmitic acid and hexitol anhydrides derived from sorbitol. It is a nonionic w/o emulsifier for neutral and mildly alkaline and acid pH systems. It is compatible with anionic, cationic and nonionic surfactants.

Applications:
LIPOSORB P is primarily used in o/w emulsions as an auxiliary emulsifier and stabilizer to balance LIPOSORB P-20.

Typical Properties:
Appearance @ 25C: Cream/tan beads
Odor: Bland, characteristic
Ionic Character: Nonionic
HLB: 6.7
Acid Value: 7.5 max.
Hydroxyl Value: 272-306
Saponification Value: 139-151
Moisture, %: 1.5 Max.

LIPO CHEMICALS INC.: Emulsifiers(Continued):

LIPOSORB S:
CTFA Adopted Name: Sorbitan Stearate

Description:
LIPOSORB S consists predominantly of the monoesters of stearic acid and hexitol anhydrides derived from sorbitol. It is a nonionic w/o emulsifier for neutral and mildly alkaline and acid pH systems. It is compatible with anionic, cationic and nonionic surfactants.

Applications:
LIPOSORB S is widely used in cosmetics, toiletries, pharmaceuticals and foods. In o/w emulsions, it functions as an auxiliary emulsifier and stabilizer to balance LIPOSORB S-20.

Typical Properties:
Appearance: Cream beads or flakes
Odor: Bland, characteristic
Ionic Character: Nonionic
HLB: 4.7
Acid Value: 5-10
Hydroxyl Value: 235-260
Saponification Value: 147-157
Moisture, %: 1.5 Max.

LIPOSORB TS:
CTFA Adopted Name: Sorbitan Tristearate

Description:
LIPOSORB TS consists predominantly of the triesters of stearic acid and hexitol anhydrides derived from sorbitol. It is a nonionic w/o emulsifier for neutral and mildly alkaline and acid pH systems. It is compatible with anionic, cationic and nonionic surfactants.

Applications:
LIPOSORB TS is utilized in cosmetics, toiletries and topical pharmaceuticals as a primary emulsifier for w/o systems. In o/w emulsions it provides effective auxiliary emulsifying and stabilizing activity when balanced with LIPOSORB S-20 or LIPOPEG 39-S.

Typical Properties:
Appearance: Tan flakes or beads
Odor: Bland, characteristic
Ionic Character: Nonionic
HLB: 2.1
Acid Value: 15 Max.
Hydroxyl Value: 65-80
Saponification Value: 175-190
Moisture, %: 1 Max.

LIPOSORB CHEMICALS INC.: Emulsifiers(Continued):

LIPOSORB O:
CTFA Adopted Name: Sorbitan Oleate

Description:
LIPOSORB O consists predominantly of the monoesters of oleic acid and hexitol anhydrides derived from sorbitol. It is a nonionic w/o emulsifier for neutral and mildly alkaline and acid pH systems. It is compatible with nonionic, anionic and cationic surfactants.

Applications:
LIPOSORB O is suggested for cosmetics, toiletries and topical pharmaceutical vehicles. It enhances the w/o activity of beeswax/borax and other w/o emulsifier systems.

Typical Properties:
  Appearance @ 25C: Yellow to amber liquid
  Odor: Bland, characteristic
  Ionic Character: Nonionic
  HLB: 4.3
  Acid Value: 8 Max.
  Hydroxyl Value: 193-210
  Saponification Value: 145-160
  Moisture, %: 1 Max.
  Conforms to NF Specifications for Sorbitan Monooleate.

LIPOSORB SQO:
CTFA Adopted Name: Sorbitan Sesquioleate

Description:
LIPOSORB SQO consists of mixed esters formed by the reaction of 1.5 moles of oleic acid with 1 mole of hexitol anhydrides derived from sorbitol. It is a nonionic w/o emulsifier for neutral and mildly alkaline and acid pH systems. It is compatible with nonionic, anionic and cationic surfactants.

Applications:
LIPOSORB SQO is a highly efficient liquid w/o emulsifier for cosmetics, toiletries and topical pharmaceuticals. It is a primary emulsifier for nonionic systems and an auxiliary emulsifier for beeswax/borax and divalent soap systems such as calcium and magnesium stearate.

Typical Properties:
  Appearance @ 25C: Yellow liquid
  Odor: Bland, characteristic
  Ionic Character: Nonionic
  HLB: 3.7
  Acid Value: 13 Max.
  Hydroxyl Value: 185-215
  Saponification Value: 145-160
  Moisture, %: 1 Max.

LIPO CHEMICALS INC.: Emulsifiers(Continued):

LIPOSORB SQO:
CTFA Adopted Name: Sorbitan Sesquioleate

Description:
LIPOSORB SQO consists of mixed esters formed by the reaction of 1.5 moles of oleic acid with 1 mole of hexitol anhydrides derived from sorbitol. It is a nonionic w/o emulsifier for neutral and mildly alkaline and acid pH systems. It is compatible with nonionic, anionic and cationic surfactants.

Applications:
LIPOSORB SQO is a highly efficient liquid w/o emulsifier for cosmetics, tolietries and topical pharmaceuticals. It is a primary emulsifier for nonionic systems and an auxiliary emulsifier for beeswax/borax and divalent soap systems such as calcium and magnesium stearate.

Typical Properties:
  Appearance @ 25C: Yellow liquid
  Odor: Bland, characteristic
  Ionic Character: Nonionic
  HLB: 3.7
  Acid Value: 13 Max.
  Hydroxyl Value: 185-215
  Saponification Value: 145-160
  Moisture, %: 1 Max.

LIPOSORB TO:
CTFA Adopted Name: Sorbitan Trioleate

Description:
LIPOSORB TO consists predominantly of the triesters of oleic acid and hexitol anhydrides derived from sorbitol. It is a nonionic w/o emulsifier for neutral and mildly alkaline and acid pH systems. It is compatible with anionic, nonionic and cationic surfactants.

Applications:
LIPOSORB TO is a primary emulsifier for nonionic w/o systems and an auxiliary emulsifier for beeswax/borax and divalent soap systems such as calcium and magnesium stearate.

Typical Properties:
  Appearance @ 25C: Amber liquid
  Odor: Bland, characteristic
  Ionic Character: Nonionic
  HLB: 1.8
  Acid Value: 13.5 Max.
  Hydroxyl Value: 55-70
  Saponification Value: 170-190
  Moisture, %: 1 Max.

LIPO CHEMICALS INC.: Emulsifiers(Continued):

LIPOSORB L-10:
CTFA Adopted Name: PEG-10 Sorbitan Laurate

Description:
LIPOSORB L-10 consists predominantly of the monolaurate esters of sorbitol and sorbitol anhydrides reacted with approximately 10 moles of ethylene oxide. It is a nonionic o/w emulsifier and solubilizer for neutral and mildly acid and alkaline pH systems. It is compatible with anionic, cationic and nonionic surfactants and has moderate electrolyte tolerance.

Applications:
LIPOSORB L-10 is a convenient fluid solubilizer for oils and fragrances in aftershaves, skin fresheners, shampoos and personal washing liquids and gels.

Typical Properties:
  Appearance @ 25C: Yellow liquid
  Odor: Bland, characteristic
  Ionic Character: Nonionic
  HLB: 14.9
  Acid Value: 2 Max.
  Hydroxyl Value: 150-170
  Saponification Value: 66-76
  Moisture, %: 3 Max.

LIPOSORB L-20:
CTFA Adopted Name: Polysorbate 20

Description:
LIPOSORB L-20 consists predominantly of the monolaurate esters of sorbitol and sorbitol anhydrides reacted with approximately 20 moles of ethylene oxide. It is a nonionic o/w emulsifier and solubilizer for neutral and mildly alkaline and acid pH systems. It is compatible with anionic, cationic and nonionic surfactants and has moderate electrolyte tolerance.

Applications:
LIPOSORB L-20 is widely used in cosmetics, toiletries, pharmaceuticals and foods.

Typical Properties:
  Appearance @ 25C: Pale yellow liquid
  Odor: Bland, characteristic
  Ionic Character: Nonionic
  HLB: 16.7
  Acid Value: 2 Max.
  Hydroxyl Value: 96-108
  Saponification Value: 40-50
  Moisture, %: 3 Max.

LIPO CHEMICALS INC.: Emulsifiers(Continued):

LIPOSORB P-20:
CTFA Adopted Name: Polysorbate 40

Description:
LIPOSORB P-20 consists predominantly of the monopalmitate esters of sorbitol and sorbitol anhydrides reacted with approximately 20 moles of ethylene oxide. It is a nonionic o/w emulsifier and solubilizer for neutral and mildly acid and alkaline pH systems. It is compatible with nonionic, anionic and cationic surfactants and has moderate electrolyte tolerance.

Applications:
LIPOSORB P-20 balanced with LIPOSORB P or LIPO GMS produces fine-textured emulsions of esters, waxes, vegetable oils and mineral oil.

Typical Properties:
  Appearance @ 25C: Yellow liquid or gel
  Odor: Bland, characteristic
  Ionic Character: Nonionic
  HLB: 15.6
  Acid Value: 2 Max.
  Hydroxyl Value: 90-107
  Saponification Value: 40-53
  Moisture, %: 3 Max.

LIPOSORB S-4:
CTFA Adopted Name: Polysorbate 61

Description:
LIPOSORB S-4 consists predominantly of the monostearate esters of sorbitol and sorbitol anhydrides reacted with approximately 4 moles of ethylene oxide. It is a nonionic o/w emulsifier for neutral and mildly alkaline or acidic pH systems. It is compatible with anionic, cationic and nonionic surfactants.

Applications:
LIPOSORB S-4 can be blended with LIPO GMS or LIPOSORB S to prepare emulsifier blends that function at HLB 6-9, the required range for petrolatum, beeswax and many triglycerides. It is suggested for skin and hair care products.

Typical Properties:
  Appearance at 25C: Tan waxy solid
  Odor: Mild, characteristic
  Ionic Character: Nonionic
  HLB: 9.6
  Acid Value: 2 Max.
  Hydroxyl Value: 170-200
  Saponification Value: 95-115
  Moisture, %: 3 Max.

LIPO CHEMICALS INC.: Emulsifiers(Continued):

LIPOSORB S-20:
CTFA Adopted Name: Polysorbate 60
Description:
LIPOSORB S-20 consists predominantly of the monstearate esters of sorbitol and sorbitol anhydrides reacted with approximately 20 moles of ethylene oxide. It is a nonionic o/w emulsifier and solubilizer for neutral, mildly alkaline and acid pH systems. It is compatible with anionic, cationic and nonionic surfactants and has moderate electrolyte tolerance.

Applications:
LIPOSORB S-20 has broad application in cosmetics, toiletries, pharmaceuticals and foods.
LIPOSORB S-20 is a functional emulsifier for pharmaceutical vehicles and a solubilizer for liquid vitamin supplements.

Typical Properties:
  Appearance @ 25C: Pale yellow liquid or gel
  Odor: Bland, characteristic
  Ionic Character: Nonionic
  HLB: 14.9
  Acid Value: 2 Max.
  Hydroxyl Value: 81-96
  Saponification Value: 45-55
  Moisture, %: 3 Max.

LIPOSORB TO-20:
CTFA Adopted Name: Polysorbate 85
Description:
LIPOSORB TO-20 consists predominantly of the trioleate esters of sorbitol and sorbitol anhydrides reacted with approximately 20 moles of ethylene oxide. It is a nonionic o/w emulsifier for neutral and mildly acid and alkaline pH systems. It is compatible with cationic, nonionic and anionic surfactants and has moderate electrolyte tolerance.

Applications:
LIPOSORB TO-20 can be blended with LIPOSORB O, LIPOSORB SQO or LIPOSORB TO to prepare fluid emulsifier systems that are highly functional at HLB 6-9.

Typical Properties:
  Appearance at 25C: Amber liquid
  Odor: Bland, characteristic
  Ionic Character: Nonionic
  HLB: 11.0
  Acid Value: 2 Max.
  Hydroxyl Value: 39-52
  Saponification Value: 82-95
  Moisture, %: 5.5 Max.

LIPO CHEMICALS INC.: Emulsifiers(Continued):

LIPOSORB O-20:
   CTFA Adopted Name: Polysorbate 80

Description:
   LIPOSORB O-20 consists prodominantly of the monooleate esters of sorbitol and sorbitol anhydrides reacted with approximately 20 moles of ethylene oxide. It is a nonionic o/w emulsifier and solubilizer for neutral and mildly alkaline and acid pH systems. It is compatible with anionic, cationic and nonionic surfactants and has moderate electrolyte tolerance.

Applications:
   LIPOSORB O-20 is a functional formulating ingredient for cosmetics, toiletries, pharmaceuticals and foods.

Typical Properties:
   Appearance @ 25C: Clear yellow liquid
   Odor: Bland, characteristic
   Ionic Character: Nonionic
   HLB: 15.0
   Acid Value: 2 Max.
   Hydroxyl Value: 65-80
   Saponification Value: 45-55
   Moisture, %: 3 Max.

LIPOSORB TS-20:
   CTFA Adopted Name: Polysorbate 65

Description:
   LIPOSORB TS-20 consists predominantly of the tristearate esters of sorbitol and sorbitol anhydrides reacted with approximately 20 moles of ethylene oxide. It is a nonionic o/w emulsifier and dispersing agent for neutral and mildly alkaline and acid pH systems. It is compatible with anionic, nonionic and cationic surfactants and has moderate electrolyte tolerance.

Applications:
   LIPOSORB TS-20 can be blended with LIPO GMS, LIPOSORB S and/or LIPOSORB TS to prepare emulsifier and dispersant systems that are highly functional at HLB 6-9.

Typical Properties:
   Appearance at 25C: Yellow to tan waxy solid
   Odor: Bland, characteristic
   Ionic Character: Nonionic
   HLB: 10.5
   Acid Value: 2 Max.
   Hydroxyl Value: 44-60
   Saponification Value: 88-98
   Moisture, %: 3 Max.

LIPO CHEMICALS INC.: Emulsifiers(Continued):

LIPOCOL L-1:
CTFA Adopted Name: Laureth-1

Description:
LIPOCOL L-1 is the ethylene glycol ether of lauryl alcohol. It is a lubricant, plasticizer, spreading and dispersing agent and nonionic auxiliary w/o emulsifier for o/w systems. It is stable over a broad pH range and is compatible with nonionic, anionic and cationic surfactants.

Applications:
LIPOCOL L-1 is suggested as a lubricant and emulsion stabilizer for o/w creams and lotions.

Typical Properties:
Appearance @ 25C: Colorless liquid
Odor: Mild, characteristic
Ionic Character: Nonionic
HLB: 3.6
Acid Value: 2 Max.
Hydroxyl Value: 231-243
Moisture, %: 0.5 Max.

LIPOCOL L-4:
CTFA Adopted Name: Laureth-4

Description:
LIPOCOL L-4 is the 4 mole ethylene oxide ether of lauryl alcohol. It is a nonionic emulsifier that has an intermediate HLB enabling it to function as an opacifier and stabilizer in o/w systems. It is stable and active over a broad pH range and has good electrolyte tolerance. It is compatible with anionic, nonionic and cationic surfactants.

Applications:
LIPOCOL L-4 is primarily used in o/w emulsions as an auxiliary emulsifier and stabilizer for LIPOCOL L-23.

Typical Properties:
Appearance @ 25C: Colorless to slightly yellow liquid
Odor: Bland, characteristic
Ionic Character: Nonionic
HLB: 9.7
Acid Value: 2 Max.
Hydroxyl Value: 145-160
Moisture, %: 0.5 Max.

LIPO CHEMICALS INC.: Emulsifiers(Continued):

LIPOCOL L-4 Special:
  CTFA Adopted Name: Laureth-4

Description:
  LIPOCOL L-4 Special is the 4 mole ethylene oxide ether of lauryl alcohol. It is a nonionic emulsifier that has an intermediate HLB enabling it to function as an opacifier and stabilizer in o/w systems. It is stable and active over a broad pH range and has good electrolyte tolerance. It is compatible with anionic, nonionic and cationic surfactants.

Applications:
  LIPOCOL L-4 Special is primarily used in o/w emulsions as an auxiliary emulsifier and stabilizer for LIPOCOL L-23.

Typical Properties:
  Appearance at 25C: Colorless liquid
  Odor: Bland, characteristic
  Ionic Character: Nonionic
  HLB: 9.7
  Acid Value: 2 Max.
  Hydroxyl Value: 145-160
  Moisture, %: 0.5 Max.

LIPOCOL L-12:
  CTFA Adopted Name: Laureth-12

Description:
  LIPOCOL L-12 is the 12 mole polyethylene glycol ether of lauryl alcohol. It is a nonionic o/w emulsifier and solubilizer suitable for use over a wide pH range. It is compatible with anionic, nonionic and cationic surfactants and has moderate electrolyte tolerance.

Applications:
  LIPOCOL L-12 is a valuable cleansing agent and solubilizer for shampoos, bubble baths, liquid detergents and personal washing preparations.

Typical Properties:
  Appearance @ 25C: White waxy solid
  Odor: Bland, characteristic
  Ionic Character: Nonionic
  HLB: 14.5
  Acid Value: 2 Max.
  Hydroxyl Value: 72-87
  Moisture, %: 1 Max.

LIPO CHEMICALS INC.: Emulsifiers(Continued):

LIPOCOL L-23:
CTFA Adopted Name: Laureth-23

Description:
LIPOCOL L-23 is the 23 mole polyethylene glycol ether of lauryl alcohol. It is a nonionic o/w emulsifier and solubilizer suitable for use over a broad pH range. It is compatible with anionic, cationic and nonionic surfactants and has good electrolyte tolerance.

Applications:
Because of its high surface activity, LIPOCOL L-23 is extensively used as a cleansing agent and solubilizer in acne scrubs, medicated shampoos, scalp treatment products and bubble baths.

Typical Properties:
   Appearance @ 25C: White solid
   Odor: Bland, characteristic
   Ionic Character: Nonionic
   HLB: 16.9
   Acid Value: 0.5 Max.
   Hydroxyl Value: 42-52
   Moisture, %: 3 Max.

LIPOCOL L-23 Special:
CTFA Adopted Name: Laureth-23

Description:
LIPOCOL L-23 Special is the 23 mole polyethylene glycol ether of lauryl alcohol. It is a nonionic o/w emulsifier and solubilizer suitable for use over a broad pH range. It is compatible with anionic, cationic and nonionic surfactants and has good electrolyte tolerance.

Applications:
Because of its high surface activity, LIPOCOL L-23 Special is extensively used as a cleansing agent and solubilizer in acne scrubs, medicated shampoos, scalp treatment products and bubble baths.

Typical Properties:
   Appearance @ 25C: White solid
   Odor: Bland, characteristic
   Ionic Character: Nonionic
   HLB: 16.9
   Acid Value: 2 Max.
   Hydroxyl Value: 42-52
   Moisture, %: 3 Max.

LIPO CHEMICALS INC.: Emulsifiers(Continued):

LIPOCOL C-2:
   CTFA Adopted Name: Ceteth-2

Description:
   LIPOCOL C-2 is the 2 mole ethylene oxide ether of cetyl alcohol. It is a nonionic w/o emulsifier that is stable and active over a broad pH range. It is compatible with anionic, nonionic and cationic surfactants.

Applications:
   LIPOCOL C-2 is primarily used in o/w emulsions as an auxiliary emulsifier and stabilizer.

Typical Properties:
   Appearance at 25C: White waxy solid
   Odor: Bland, characteristic
   Ionic Character: Nonionic
   HLB: 5.3
   Acid Value: 1 Max.
   Hydroxyl Value: 160-180
   Moisture, %: 1 Max.

LIPOCOL C-10:
   CTFA Adopted Name: Ceteth-10

Description:
   LIPOCOL C-10 is the 10 mole ethylene oxide ether of cetyl alcohol. It is a nonionic o/w emulsifier that is stable and active over a broad pH range. It is compatible with anionic, nonionic and cationic surfactants and has moderate electrolyte tolerance.

Applications:
   LIPOCOL C-10 forms effective nonionic emulsifier systems when balanced with LIPOCOL C-2 and/or LIPOCOL C-20.

Typical Properties:
   Appearance @ 25C: White, waxy solid
   Odor: Bland, characteristic
   Ionic Character: Nonionic
   HLB: 12.9
   Acid Value: 1 Max.
   Hydroxyl Value: 75-90
   Moisture, %: 3 Max.

LIPO CHEMICALS INC.: Emulsifiers(Continued):

LIPOCOL C-20:

Description:
LIPOCOL C-20 is the 20 mole ethylene oxide ether of cetyl alcohol. It is a nonionic o/w emulsifier that is stable and active over a broad pH range. It is compatible with cationic, nonionic and anionic surfactants and has moderate electrolyte tolerance.

Applications:
Emulsifier systems formed by balancing LIPOCOL C-20 with LIPOCOL C-2, LIPOCOL C or LIPOCOL S offer wide latitude in the preparation of fine-textured creams and lotions for skin and hair care.

Typical Properties:
   Appearance at 25C: White waxy solid
   Odor: Bland, characteristic
   Ionic character: Nonionic
   HLB: 15.7
   Acid Value: 2 Max.
   Hydroxyl Value: 45-60
   Moisture, %: 3 max.

LIPOCOL S-2:
   CTFA Adopted Name: Steareth-2

Description:
LIPOCOL S-2 is the 2 mole ethylene oxide ether of stearyl alcohol. It is a nonionic w/o emulsifier that is stable and active over a broad pH range. It is compatible with cationic, nonionic and anionic surfactants

Applications:
LIPOCOL S-2 is primarily used as an auxiliary emulsifier and stabilizer in o/w emulsions. Effective nonionic emulsifier systems can be formed by balancing LIPOCOL S-20.

Typical Properties:
   Appearance at 25C: White to translucent waxy solid
   Odor: Bland, characteristic
   Ionic character: Nonionic
   HLB: 4.9
   Acid Value: 1 Max.
   Hydroxyl Value: 155-165
   Moisture, %: 1 Max.

LIPO CHEMICALS INC.: Emulsifiers(Continued):

LIPOCOL S-10:
    CTFA Adopted Name: Steareth-10

Description:
    LIPOCOL S-10 is the 10 mole ethylene oxide ether of stearyl alcohol. It is a nonionic o/w emulsifier that is active and stable over a broad pH range. It is compatible with cationic, nonionic and anionic surfactants and has moderate electrolyte tolerance.

Applications:
    LIPOCOL S-10 combined with LIPOCOL S-2 and/or LIPOCOL S-20 form nonionic ester-free emulsifier systems.

Typical Properties:
    Appearance at 25C: White waxy solid
    Odor: Bland, characteristic
    Ionic Character: Nonionic
    HLB: 12.4
    Acid Value: 1 Max.
    Hydroxyl Value: 75-90
    Moisture, %: 3 Max.

LIPOCOL S-20:
    CTFA Adopted Name: Steareth-20

Description:
    LIPOCOL S-20 is the 20 mole ethylene oxide ether of stearyl alcohol. It is a nonionic o/w emulsifier that is stable and active over a broad pH range. It is compatible with anionic, cationic and nonionic surfactants and has moderate electrolyte tolerance.

Applications:
    LIPOCOL S-20 forms effective emulsifier systems when balanced with LIPOCOL S-2, LIPOCOL S or LIPOCOL C.

Typical Properties:
    Appearance @ 25C: White waxy solid
    Odor: Bland, characteristic
    Ionic Character: Nonionic
    HLB: 15.3
    Acid Value: 1 Max.
    Hydroxyl Value: 45-60
    Moisture, %: 3 Max.

LIPO CHEMICALS INC.: Emulsifiers(Continued):

LIPOCOL O-2:
CTFA Adopted Name: Oleth-2

Description:
LIPOCOL O-2 is the 2 mole ethylene oxide ether of oleyl alcohol. It is a nonionic w/o emulsifier that is stable and active over a broad pH range. It is compatible with anionic, nonionic and cationic surfactants.

Applications:
LIPOCOL O-2 is a primary w/o emulsifier for nonionic systems and an auxiliary emulsifier for beeswax/borax and divalent soap systems such as calcium and magnesium stearate. When balanced with the LIPOCOL O-20, it produces fine-textured nonionic o/w creams and lotions.

Typical Properties:
Appearance @ 25C: Yellow oily liquid
Odor: Bland, characteristic
Ionic Character: Nonionic

LIPOCOL O-5:
CTFA Adopted Name: Oleth-5

Description:
LIPOCOL O-5 is the 5 mole ethylene oxide ether of oleyl alcohol. It is a nonionic w/o emulsifier that is stable and active over a broad pH range. It is compatible with anionic, nonionic and cationic surfactants.

Applications:
LIPOCOL O-5 is a primary w/o emulsifier for nonionic systems. When balanced with the LIPOCOL O-20, it produces fine-textured nonionic o/w creams and lotions.

Typical Properties:
Appearance @ 25C: Yellow semi solid
Odor: Bland, characteristic
Ionic Character: Nonionic
HLB: 8.8
Acid Value: 1 Max.
Hydroxyl Value: 120-135
Moisture, %: 1 Max.

LIPO CHEMICALS INC.: Emulsifiers(Continued):

LIPOCOL O-20:
   CTFA Adopted Name: Oleth-20

Description:
   LIPOCOL O-20 is the 20 mole ethylene oxide adduct of oleyl alcohol. It is a nonionic o/w emulsifier and solubilizer. It is compatible with anionic, nonionic and cationic surfactants, has moderate electrolyte tolerance and is stable over the pH range normally encountered in cosmetics and toiletries.

Applications:
   LIPOCOL O-20 forms effective emulsifier systems when balanced with LIPOCOL O-2 or LIPO GMS 450. It is also a solubilizer for fragrances and other polar oils.

Typical Properties:
   Appearance at 25C: Off-white waxy solid
   Odor: Bland, characteristic
   Ionic Character: Nonionic
   HLB: 15.3
   Acid Value: 2 Max.
   Hydroxyl Value: 45-65
   Moisture, %: 1 Max.

LIPOCOL SC-4:
   CTFA Adopted Name: Ceteareth-4

Description:
   LIPOCOL SC-4 is the 4 mole ethylene oxide ether of stearyl/cetyl alcohol. It is a nonionic w/o emulsifier that is stable and active over a broad pH range. It is compatible with cationic, anionic and nonionic surfactants.

Applications:
   LIPOCOL SC-4 is principally used as an auxiliary emulsifier and stabilizer in o/w emulsions. It is especially valuable in hair rinses and conditioners based on quaternary ammonium compounds where it provides opacity, increased body and lubricity.

Typical Properties:
   Appearance @ 25C: White waxy solid
   Odor: Bland, characteristic
   Ionic Character: Nonionic
   HLB: 8.0
   Acid Value: 1 Max.
   Hydroxyl Value: 130-150
   Moisture, %: 1 max.

LIPO CHEMICALS INC.: Emulsifiers(Continued):

LIPOCOL SC-15:
CTFA Adopted Name: Ceteareth-15

Description:
LIPOCOL SC-15 is the 15 mole ethylene oxide ether of stearyl/cetyl alcohol. It is a nonionic o/w emulsifier that is stable and active over a broad pH range. It is compatible with cationic, anionic and nonionic surfactants and has good electrolyte tolerance.

Applications:
LIPOCOL SC-15 is a primary emulsifier for o/w systems. It is particularly effective when balanced with LIPOCOL SC-4.

Typical Properties:
Appearance @ 25C: White waxy solid
Odor: Bland, characteristic
Ionic Character: Nonionic
HLB: 14.3
Acid Value: 2 Max.
Hydroxyl Value: 50-65
Moisture, %: 1 Max.

LIPOCOL SC-20:
CTFA Adopted Name: Ceteareth-20

Description:
LIPOCOL SC-20 is the 20 mole ethylene oxide ether of stearyl/cetyl alcohol. It is a nonionic o/w emulsifier that is stable over a broad pH range. It has good electrolyte tolerance and is compatible with cationic, anionic and nonionic surfactants.

Applications:
LIPOCOL SC-20 is a primary o/w emulsifier and solubilizer. It is suggested for bubble baths, shampoos and personal washing liquids.

Typical Properties:
Appearance @ 25C: White waxy solid
Odor: Mild, characteristic
Ionic Character: Nonionic
HLB: 15.5
Acid Value: 1 Max.
Hydroxyl Value: 45-60
Moisture, %: 3.0 Max.

LIPO CHEMICALS INC.: Emulsifiers(Continued):

LIPOCOL M-4:
  Myreth-4

Description:
  LIPOCOL M-4 consists essentially of the 4 mole ethylene oxide adduct of myristyl alcohol. It is a nonionic emulsifier with an intermediate HLB. It is stable and active over a broad pH range and is compatible with anionic, nonionic and cationic surfactants.

Applications:
  LIPOCOL M-4 is suggested as an auxiliary emulsifier and stabilizer for o/w emulsions and as a dispersing agent for bath/body oils.

Typical Properties:
  Appearance @ 25C: Essentially colorless liquid
  Odor: Mild, characteristic
  Ionic Character: Nonionic
  HLB: 8.8
  Acid Value: 2 Max.
  Hydroxyl Value: 140-160

LIPOCOL TD-6:
  CTFA Adopted Name: Trideceth-6

Description:
  LIPOCOL TD-6 is the 6 mole ethoxylate of tridecyl alcohol. It is a wetting agent and a nonionic emulsifier that has an intermediate HLB enabling it to function as an opacifier and stabilizer in o/w systems. It is stable and active over a broad pH range and is compatible with anionic, nonionic and cationic surfactants.

Applications:
  LIPOCOL TD-6 is suggested as an opacifier and auxiliary emulsifier for o/w systems, a wetting agent and a solubilizer for detergent systems.

Typical Properties:
  Appearance @ 25C: White liquid to solid
  Odor: Mild, characteristic
  Ionic Character: Nonionic
  HLB: 11.4
  Acid Value: 1 Max.
  Hydroxyl Value: 115-135
  Moisture, %: 1 Max.

LIPO CHEMICALS INC.: Emulsifiers(Continued):

LIPOCOL TD-12:
  CTFA Adopted Name: Trideceth-12

Description:
  LIPOCOL TD-12 is the 12 mole ethoxylate of tridecyl alcohol. It is a nonionic o/w emulsifier and solubilizer. It is compatible with anionic, nonionic and cationic surfactants and is stable and active over a broad pH range.

Applications:
  LIPOCOL TD-12 is suggested as a primary emulsifier for o/w systems and as a solubilizer for fragrances and other polar compounds in soaps and detergents.

Typical Properties:
  Appearance @ 25C: White semi-solid
  Odor: Mild, characteristic
  Ionic Character: Nonionic
  HLB: 14.6
  Acid Value: 1 Max.
  Hydroxyl Value: 70-85
  Moisture, %: 1 Max.

LIPOPEG 4-L:
  CTFA Adopted Name: PEG-8 Laurate

Description:
  LIPOPEG 4-L consists essentially of the 8 mole polyethylene glycol monoester of lauric acid. It is a nonionic o/w emulsifier for neutral and mildly acidic and alkaline systems. It is compatible with anionic, nonionic and cationic surfactants and has moderate electrolyte tolerance.

Applications:
  LIPOPEG 4-L is a convenient fluid emulsifier for skin and hair care creams and lotions and emulsified makeups. It is a thickening agent for shampoos and creme rinses.

Typical Properties:
  Appearance @ 25C: Yellow liquid
  Odor: Mild, characteristic
  Ionic Character: Nonionic
  HLB: 13.9
  Acid Value: 5 Max.
  Saponification Value: 85-105
  Moisture, %: 3 Max.

LIPO CHEMICALS INC.: Emulsifiers(Continued):

LIPOPEG 4-S:
CTFA Adopted Name: PEG-8 Stearate

Description:
LIPOPEG 4-S is the 8 mole polyethylene glycol ester of stearic acid. It is a nonionic o/w emulsifier for neutral and mildly alkaline and acid pH systems. It is compatible with anionic, cationic and nonionic surfactants and has limited electrolyte tolerance.

Applications:
LIPOPEG 4-S is a versatile general-purpose emulsifier that produces rich, full bodied creams and lotions. It can be used as the sole emulsifier for systems having required HLB values of 10 to 12.

Typical Properties:
    Appearance @ 25C: White soft solid
    Odor: Bland, characteristic
    Ionic Character: Nonionic
    HLB: 11.2
    Acid Value: 5 Max.
    Saponification Value: 80-90
    Moisture, %: 1 Max.

LIPOPEG 15-S:
CTFA Adopted Name: PEG-6-32 Stearate

Description:
LIPOPEG 15-S is an ester of stearic acid and a polyethylene glycol of approximately 30 moles. It is an excellent nonionic o/w emulsifier for mildly alkaline, neutral and acid pH systems. It is compatible with anionic, nonionic and cationic surfactants and has very good electrolyte tolerance.

Applications:
LIPOPEG 15-S is a versatile emulsifier for cosmetics, toiletries and topical pharmaceuticals. It is effective in formulations containing high concentrations of electrolytes or high fragrance levels. It is suggested where maximum wetting, dispersing or solubilizing are required.

Typical Properties:
    Appearance: Ivory waxy solid
    Odor: Bland, characteristic
    Ionic Character: Nonionic
    HLB: 15-16
    Acid Value: 5 Max.
    Saponification Value: 25-35
    Hydroxyl Value: 25-40
    Moisture, %: 3 Max.

LIPO CHEMICALS INC.: Emulsifiers(Continued):

LIPOPEG 39-S:
CTFA Adopted Name: PEG-40 Stearate

Description:
LIPOPEG 39-S is the 40 mole polyethylene glycol ester of stearic acid. It is a powerful nonionic o/w emulsifier for mildly alkaline, neutral and acid pH systems. It is compatible with anionic, nonionic and cationic surfactants and has very good electrolyte tolerance.

Applications:
LIPOPEG 39-S is a versatile emulsifier for cosmetics, toiletries and topical pharmaceuticals. When balanced with LIPO GMS or LIPOSORB TS it produces fine-textured emulsions of oils, fats and waxes.

Typical Properties:
Appearance: Ivory waxy solid or flake
Odor: Bland, characteristic
Ionic Character: Nonionic
HLB: 16.9
Acid Value: 2 Max.
Saponification Value: 25-35
Hydroxyl Value: 25-40
Moisture, %: 3 max.

LIPOPEG 6000-DS:
PEG-150 Distearate

Description:
LIPOPEG 6000-DS consists essentially of the 150 mole polyoxyethylene diester of stearic acid. It is a nonionic o/w emulsifier for neutral and mildly acidic and alkaline systems. It is compatible with anionic, nonionic and cationic surfactants and has good electrolyte tolerance.

Applications:
LIPOPEG 6000-DS is an o/w emulsifier for creams and lotions, especially for systems containing high concentrations of electrolytes. It is an effective thickener for transparent shampoos and personal washing preparations.

Typical Properties:
Appearance @ 25C: White to off-white flakes
Odor: Mild, characteristic
Ionic Character: Nonionic
HLB: 18.4
Acid Value: 10 Max.
Saponification Value: 12-20

LIPO CHEMICALS INC.: Emulsifiers(Continued):

LIPOPEG 6000-DL:
CTFA Adopted Name: PEG-150 Dilaurate

Description:
LIPOPEG 6000-DL consists essentially of the 150 mole polyethylene glycol diester of lauric acid. It is a nonionic o/w emulsifier for neutral and mildly acidic and alkaline systems. It is compatible with anionic, nonionic and cationic surfactants and has good electrolyte tolerance.

Applications:
LIPOPEG 6000-DL is a powerful o/w emulsifier for creams and lotions. Because it remains functional in systems containing high concentrations of polyvalent electrolytes, it is particularly suggested for antiperspirant and acid-mantle skin and hair care products.

Typical Properties:
Appearance @ 25C: White waxy solid or flakes
Odor: Mild, characteristic
Ionic Character: Nonionic
HLB: 18.7
Acid Value: 10 Max.
Saponification Value: 12-24

LIPOCOL B:
CTFA Adopted Name: PEG-9 Stearate (and) PEG-9 Laurate (and) PEG-2 Laurate SE

Description:
LIPOCOL B is a blend of polyethylene glycol esters of stearic acid and lauric acid designed to provide opacity and thickening in o/w systems. In contrast to self-emulsifying waxes and absorption bases, LIPOCOL B is entirely surface active and contains no oils, fats, fatty alcohols, waxes or petrolatum. This o/w emulsifier has a slight anionic charge and is effective in mildly alkaline pH systems. LIPOCOL B is compatible with nonionic and anionic surfactants and has good electrolyte tolerance.

Applications:
LIPOCOL B is especially suggested as an opacifier and thickener for formulations such as bromate neutralizers.

Typical Properties:
Appearance @ 25C: White to off-white soft waxy solid
Odor: Fatty, characteristic
Ionic Character: Anionic
HLB: 11.0
Acid Value: 5 Max.
Saponification Value: 94-104
Moisture, %: 1 Max.

LIPO CHEMICALS INC.: Emulsifiers(Continued):

LIPOPEG 2-DL:
CTFA Adopted Name: PEG-4 Dilaurate

Description:
LIPOPEG 2-DL consists essentially of the 4 mole polyethylene glycol diester of lauric acid. It is a nonionic o/w emulsifier for neutral and mildly alkaline and acidic systems. It is compatible with nonionic, anionic and cationic surfactants.

Applications:
LIPOPEG 2-DL is a convenient fluid emulsifier, dispersant and spreading agent. It is suggested for creams, lotions, hair dressings, bath and body oils.

Typical Properties:
  Appearance at 25C: Yellow liquid
  Odor: Mild, characteristic
  Ionic Character: Nonionic
  HLB: 6
  Acid Value: 10 Max.
  Saponification Value: 170-190

LIPOPEG 4-DL:
CTFA Adopted Name: PEG-8 Dilaurate

Description:
LIPOPEG 4-DL consists essentially of the 8 mole polyethylene glycol diester of lauric acid. It is a nonionic o/w emulsifier for neutral and mildly alkaline and acidic systems. It is compatible with anionic, nonionic and cationic surfactants and has moderate electrolyte tolerance.

Applications:
LIPOPEG 4-DL is a convenient fluid emulsifier for skin and hair care creams and lotions. It is a spreading and dispersing agent for bath and body oils.

Typical Properties:
  Appearance @ 25C: Yellow liquid
  Odor: Bland, characteristic
  Ionic Character: Nonionic
  HLB: 10
  Acid Value: 10 Max.
  Saponification Value: 125-142
  Moisture, %: 1 max.

LIPO CHEMICALS INC.: Emulsifiers(Continued):

LIPOPEG 4-DO:
CTFA Adopted Name: PEG-8 Dioleate

Description:
LIPOPEG 4-DO consists essentially of the 8 miole polyethylene glycol diester of oleic acid. It is a nonionic w/o emulsifier and dispersant. It is stable at neutral and mildly acidic and alkaline pH levels and is compatible with nonionic, anionic and cationic surfactants.

Applications:
LIPOPEG 4-DO is suggested for skin and hair care creams and lotions and for anhydrous and emulsified makeups. It is a spreading and dispersing agent for bath and body oils.

Typical Properties:
Appearance @ 25C: Yellow liquid
Odor: Mild, slightly fatty
Ionic Character: Nonionic
HLB: 8
Acid Value: 10 Max.
Saponification Value: 113-128

LIPOPEG 4-DS:
CTFA Adopted Name: PEG-8 Distearate

Description:
LIPOPEG 4-DS is the 8 mole polyethylene glycol diester of stearic acid. It is a nonionic W/O emulsifier for neutral and mildly alkaline and acid pH systems. It is compatible with anionic, cationic and nonionic surfactants.

Applications:
LIPOPEG 4-DS is a good thickener in cosmetic formulations. It is recommended for use in hair care products, deodorant and antiperspirant sticks, creams, lotions and cream shampoos as a thickener or viscosity modifier.

Typical Properties:
Appearance @ 25C: White to tan solid wax
Odor: Bland, characteristic
Ionic Character: Nonionic
HLB: 8
Acid Value: 10 Max.
Saponification Value: 113-128

LIPO CHEMICALS INC.: Emulsifiers(Continued):

LIPOPEG 2-L:
    CTFA Adopted Name: PEG-4 Laurate

Description:
    LIPOPEG 2-L consists essentially of the 4 mole polyethylene glycol monoester of lauric acid. It is a nonionic o/w emulsifier for neutral and mildly acidic and alkaline systems. It is compatible with anionic, nonionic and cationic surfactants and has poor to moderate electrolyte tolerance.

Applications:
    LIPOPEG 2-L is a convenient fluid emulsifier and dispersant for bath oils, creams and lotions. It is also used as a thickening agent for shampoos and creme rinses.

Typical Properties:
    Appearance @ 25C: Clear, light yellow liquid
    Odor: Mild, characteristic
    HLB: 9
    Acid Value: 5 Max.
    Saponification Value: 140-160
    Iodine Value: 10 Max.

LIPOPEG 4-O:
    CTFA Adopted Name: PEG-8 Oleate

Description:
    LIPOPEG 4-O consists essentially of the 8 mole polyethylene glycol monoester of oleic acid. It is a nonionic o/w emulsifier for neutral and mildly acidic and alkaline pH systems. It is compatible with anionic, cationic and nonionic surfactants and has limited electrolyte tolerance.

Applications:
    LIPOPEG 4-O is a convenient fluid emulsifier for o/w creams and lotions. It is suggested for skin and hair care preparations and emulsified makeups.

Typical Properties:
    Apprearance @ 25C: Yellow to amber oil
    Odor: Mild, slightly fatty
    Ionic Character: Nonionic
    HLB: 11
    Acid Value: 5 Max.
    Saponification Value: 80-90

LIPO CHEMICALS INC.: Emulsifiers(Continued):

LIPO EGMS:
  CTFA Adopted Name: Glycol Stearate

LIPO EGDS:
  CTFA Adopted Name: Glycol Distearate

Description:
  LIPO EGMS and LIPO EGDS are esters of ethylene glycol stearic acid. LIPO EGMS is primarily monoester and LIPO EGDS is primarily diester. Both are neutral nonionic w/o emulsifiers and stabilizers for o/w systems. They are compatible with nonionic, anionic and cationic surfactants.
Applications:
  LIPO EGMS and LIPO EGDS are suggested as opacifiers, thickeners and pearlescing agents for creams, lotions, shampoos, personal washing preparations and hand dishwashing and laundry products.
Typical Properties:
LIPO EGMS:
  Appearance @25C: White to off-white flakes or beads
  Odor: Bland, slightly fatty
  Ionic Character: Nonionic
  HLB: 2
  Acid Value: 6 Max.
  Saponification Value: 175-190
LIPO EGDS:
  Appearance @ 25C: White to off-white flakes or beads
  Odor: Bland, slightly fatty
  Ionic Character: Nonionic
  HLB: 1
  Acid Value: 7 Max.
  Saponification Value: 190-205

LIPO DGLS:
  CTFA Adopted Name: PEG-2 Laurate SE

Description:
  LIPO DGLS consists essentially of the ester of lauric acid and diethylene glycol. It has been made water-dispersible by the incorporation of a small amount of potassium laurate. It is an anionic o/w emulsifier for neutral or slightly alkaline systems. It is compatible with anionic and nonionic surfactants.
Applications:
  LIPO DGLS is a dispersing agent for bath oils and a convenient fluid emulsifier for cosmetic and toiletry skin and hair care lotions.
Typical Properties:
  Appearance @ 25C: Yellow to amber liquid
  Odor: Bland, characteristic
  Ionic Character: Anionic
  HLB: 7.4
  Acid Value: 4 Max.
  Saponification Value: 160-170

LIPO CHEMICALS INC.: Emulsifiers(Continued):

LIPO GMS 450:
CTFA Adopted Name: Glyceryl Stearate

Description:
LIPO GMS 450 consists of the mixed esters of glycerine and stearic acid. It contains a minimum of 40% monoester. It is a neutral nonionic w/o emulsifier and stabilizer for o/w systems. It is stable and active over a broad pH range and is compatible with anionic, nonionic and cationic surfactants.

Applications:
LIPO GMS 450 is widely used in cosmetics, toiletries and pharmaceuticals primarily as a consistency builder and emulsion stabilizer for o/w creams and lotions.

LIPO GMS 450 is an emulsifier, plasticizer and conditioner for chewing gums, caramels, coffee whiteners, whipped toppings, shortenings and baked goods.

Typical Properties:
Appearance @ 25C: White to off-white beads or flakes
Odor: Bland, characteristic
Ionic Character: Nonionic
HLB: 2.8
Monoester, %: 40 Min.
Free glycerine, %: 1 Max.
Melting Point, C: 56-64
Acid Value: 5 Max.

LIPO GMS 600:
CTFA Adopted Name: Glyceryl Stearate

Description:
LIPO GMS 600 consists of the mixed esters of glycerine and stearic acid. It contains a minimum of 60% monoester. It is a neutral nonionic w/o emulsifier and stabilizer for o/w systems. It is stable and active over a broad pH range and is compatible with anionic, nonionic and cationic surfactants.

Applications:
LIPO GMS 600 is widely used in cosmetics, toiletries and pharmaceuticals primarily as a consistency builder and emulsion stabilizer for o/w creams and lotions.

Typical Properties:
Appearance @ 25C: White to off-white beads or flakes
Odor: Bland, characteristic
Ionic Character: Nonionic
HLB: 3.2
Monoester, %: 60 Min.
Free Glycerine, %: 1.5 Max.
Melting Point, C: 60-69
Acid Value: 5 Max.

LIPO CHEMICALS INC.: Emulsifiers(Continued):

LIPO GMS 470:
Glyceryl Stearate SE

Description:
LIPO GMS 470 consists of esters of glycerine and stearic acid made self-emulsifying by the incorporation of a small amount of potassium stearate. It is an anionic o/w emulsifier and opacifier for neutral and alkaline pH systems. It is compatible with anionic and nonionic surfactants.

Applications:
LIPO GMS 470 disperses readily in hot water to form fine-textured emulsions.

Typical Properties:
Appearance @ 25C: White to off-white beads or flakes
Odor: Bland, characteristic
Ionic Character: Anionic
HLB: 5
Monoester, %: 30 Min.
Free Glycerine, %: 10 Max.
Melting Point, C: 55-61
Acid Value: 5 Max.

LIPOMULSE 165:
CTFA Adopted Name: Glyceryl Stearate (and) PEG-100 Stearate

Description:
LIPOMULSE 165 is a blend of glyceryl stearate and PEG-100 stearate. It is a nonionic o/w emulsifier and viscosity builder for mildly alkaline, neutral and acid pH systems. It is compatible with anionic, cationic and nonionic surfactants and has exceptional electrolyte tolerance.

Applications:
LIPOMULSE 165 is a general purpose self-emulsifying glyceryl monostearate for cosmetics, toiletries and topical pharmaceuticals.
LIPOMULSE 165 is an especially valuable emulsifier for neutral or acid mantle creams and lotions and for antiperspirants or other products containing high concentrations of electrolytes.

Typical Properties:
Appearance @ 25C: White beads or flake
Odor: Bland, characteristic
Ionic Character: Nonionic
HLB: 11
Acid Value: 2 Max.
Iodine Value: 3 Max.
Saponification Value: 90-100
Moisture, %: 2 Max.

LIPO CHEMICALS INC.: Emulsifiers(Continued):

LIPOLAN 31:
CTFA Adopted Name: PEG-24 Hydrogenated Lanolin

Description:
LIPOLAN 31 is the 24 mole polyethylene glycol ether of the lanolin alcohol complex formed by the controlled hydrogenation of lanolin. It is a nonionic o/w emulsifier and solubilizer suitable for use over a broad pH range. It is compatible with anionic, nonionic and cationic surfactants and has good electrolyte tolerance.

Applications:
LIPOLAN 31 is a versatile formulating ingredient for cosmetics, toiletries and topical pharmaceuticals.

Typical Properties:
Appearance @ 25C: Cream colored waxy solid
Odor: Bland, characteristic
Acid Value: 2 Max.
Hydroxyl Value: 35-55
Saponification Value: 8 Max.
Moisture, %: 2 Max.

LIPOWAX P:
Cetearyl Alcohol (and) Polysorbate 60

Description:
LIPOWAX P is a balanced blend of lipophilic high molecular weight fatty alcohols and hydrophilic ethoxylated products. It is a nonionic o/w self-emulsifying wax for neutral and mildly acidic and alkaline pH systems. It is compatible with anionic, nonionic and cationic surfactants and has moderate electrolyte tolerance.

Applications:
LIPOWAX P disperses easily in warm water to form elegant opaque emulsions. Consistency can be varied simply by adjusting the concentration.

Typical Properties:
Appearance @ 25C: Creamy white waxy solid or flake
Odor: Bland, characteristic
Ionic Character: Nonionic
Hydroxyl Value: 178-192
Iodine Value: 3.5 Max.
Saponification Value: 14 Max.
Melting Point, C: 48-52
pH, 3% aq. dispersion: 5.5-7.0

LIPO CHEMICALS INC.: Emulsifiers(Continued):

LIPOWAX D:
    Cetearyl Alcohol (and) Ceteareth-20

Description:
    LIPOWAX D consists of high molecular weight saturated fatty alcohols and their ethylene oxide adducts in ratios designed to provide optimum emulsification and consistency development. This nonionic o/w self-emulsifying wax is compatible with anionic, cationic and nonionic surfactants, has good electrolyte tolerance and is stable and active over the pH range normally encountered in cosmetics and toiletries.

Applications:
    LIPOWAX D readily disperses in hot water to form fine-textured opaque emulsions.

Typical Properties:
    Appearance @ 25C: White to off-white waxy solid or flakes
    Odor: Bland, characteristic
    Ionic Character: Nonionic
    HLB: 11
    Acid Value: 1 Max.
    Iodine Value: 2 Max.
    Saponification Value: 5 Max.
    Moisture, %: 1 Max.
    Melting Range, C: 46-55

LIPOWAX G:
    CTFA Adopted Name: Stearyl Alcohol (and) Ceteareth-20

Description:
    LIPOWAX G consists of stearyl alcohol and ethoxylated cetyl/stearyl alcohol blended to provide optimum emulsification for fluid systems. This nonionic o/w self-emulsifying wax is compatible with anionic, cationic and nonionic surfactants, has good electrolyte tolerance and is stable and active over the pH range normally encountered in cosmetics and toiletries.

Applications:
    LIPOWAX G disperses readily in hot water to form fine-textured opaque emulsions of low viscosity.

Typical Properties:
    Appearance @ 25C: White to off-white waxy solid or flake
    Odor: Bland, characteristic
    Ionic Character: Nonionic
    Acid Value: 1 Max.
    Saponification Value: 1 Max.
    Melting Range, C: 55-63

LIPO CHEMICALS INC.: Emulsifiers(Continued):

LIPOWAX P-31:
   Emulsifying Wax

Description:
   LIPOWAX P-31 is a balanced blend of lipophilic high molecular weight fatty alcohols and hydrophilic ethoxylated products. It is a nonionic o/w self-emulsifying wax for neutral and mildly acidic and alkaline pH systems. It is compatible with anionic and cationic surfactants and has moderate electrolyte tolerance.

Applications:
   LIPOWAX P-31 disperses easily in warm water to form elegant opaque emulsions. Consistency can be varied by adjusting the concentration.

Typical Properties:
   Appearance @ 25C: Creamy white waxy solid or flake
   Odor: Bland, characteristic
   Ionic Character: Nonionic
   Hydroxyl Value: 178-192
   Iodine Value: 3.5 Max.
   Saponification Value: 14 Max.
   Melting Point, C: 48-52
   pH, 3% aq. dispersion: 5.5-7.0

LIPOWAX PR:
   Cetearyl Alcohol (and) Polysorbate 60 (and) PEG-150 Stearate (and) Steareth-20

Description:
   LIPOWAX PR is a balanced blend of lipophilic high molecular weight fatty alcohols and hydrophilic ethoxylated products. It is a nonionic o/w self-emulsifying wax for neutral and mildly acidic and alkaline pH systems. It is compatible with anionic, nonionic and cationic surfactants and has moderate electrolyte tolerance.

Applications:
   LIPOWAX PR disperses easily in warm water to form elegant opaque emulsions. Consistency can be varied simply by adjusting the concentration.

Typical Properties:
   Appearance @ 25C: Creamy white waxy solid or flake
   Odor: Bland, characteristic
   Ionic Character: Nonionic
   Hydroxyl Value: 178-192
   Iodine Value: 3 Max.
   Saponification Value: 14 Max.
   Melting Point, C: 48-52

LIPO CHEMICALS INC.: Emulsifiers(Continued):

LIPOWAX P-NATURAL:
Emulsifying Wax NF

Description:
LIPOWAX P-Natural is a balanced blend of lipophilic high molecular weight fatty alcohols and hydrophilic ethoxylated products derived from natural ingredients. It is a nonionic o/w self-emulsifying wax for neutral and mildly acidic and alkaline pH systems. It is compatible with anionic, nonionic and cationic surfactants and has moderate electrolyte tolerance.

Applications:
LIPOWAX P - Natural disperses easily in warm water to form elegant opaque emulsions.

Typical Properties:
Appearance @ 25C: Creamy white waxy solid or flake
Odor: Bland, characteristic
Ionic Character: Nonionic
Hydroxyl Value: 178-192
Iodine Value: 3.5 Max.
Saponification Value: 14 Max.
Melting Point, C: 48-52
pH, 3% aq. dispersion: 5.5-7.0

LIPAMIDE S:
CTFA Adopted Name: Stearamide DEA

Description:
LIPAMIDE S is the 1:1 diethanolamide of stearic acid. It is a nonionic o/w emulsifier, thickener, emulsion stabilizer and lubricant. It is stable and active over a broad pH range and has excellent electrolyte tolerance. It is compatible with anionic, nonionic and cationic surfactants.

Applications:
LIPAMIDE S disperses readily in hot water. At 1% to 5%, increasingly viscous thixotropic lotions are formed.

Typical Properties:
Appearance @ 25C: Off-white wax
Odor: Mild, waxy
Acid Value: 12-18
pH, 1% aq.: 8.5-10.5
Melting Point, C: 41-46

LONZA INC.: LONZEST Sorbitan Esters:

Emulsifier, solubilizer and stabilizer with low toxicity and FDA Approval. Used in a variety of product/markets, especially foods, personal care products and industrial applications:

Product:

LONZEST SML:
  CTFA Designation: Sorbitan Laurate
  Physical Form: Liquid
  Ionic Type: Nonionic
  % Active: 100

LONZEST SMO:
  CTFA Designation: Sorbitan Oleate
  Physical Form: Liquid
  Ionic Type: Nonionic
  % Active: 100

LONZEST SMP:
  CTFA Designation: Sorbitan Palmitate
  Physical Form: Solid
  Ionic Type: Nonionic
  % Active: 100

LONZEST SMS:
  CTFA Designation: Sorbitan Stearate
  Physical Form: Flake
  Ionic Type: Nonionic
  % Active: 100

LONZEST STO:
  CTFA Designation: Sorbitan Trioleate
  Physical Form: Liquid
  Ionic Type: Nonionic
  % Active: 100

LONZEST STS:
  CTFA Designation: Sorbitan Tristearate
  Physical Form: Solid
  Ionic Type: Nonionic Type
  % Active: 100

LONZEST SML-20:
  CTFA Designation: Polysorbate 20
  Physical Form: Liquid
  Ionic Type: Nonionic
  % Active: 100

LONZA INC.: LONZEST Sorbitan Esters(Continued):

Product:

LONZEST SMO-5:
    CTFA Designation: Polysorbate 81
    Physical Form: Liquid
    Ionic Type: Nonionic
    % Active: 100

LONZEST SMO-20:
    CTFA Designation: Polysorbate 80
    Physical Form: Liquid
    Ionic Type: Nonionic
    % Active: 100

LONZEST SMP-20:
    CTFA Designation: Polysorbate 40
    Physical Form: Liquid
    Ionic Type: Nonionic
    % Active: 100

LONZEST SMS-20:
    CTFA Designation: Polysorbate 60
    Physical Form: Liquid
    Ionic Type: Nonionic
    % Active: 100

LONZEST STO-20:
    CTFA Designation: Polysorbate 85
    Physical Form: Liquid
    Ionic Type: Nonionic
    % Active: 100

LONZEST STS-20:
    CTFA Designation: Polysorbate 65
    Physical Form: Solid
    Ionic Type: Nonionic
    % Active: 100

MAYCO OIL & CHEMICAL CO.: EMULAMID Lubricant-Emulsifiers:

EMULAMID AD-50 Lubricant-Emulsifier:

Typical Inspections:
  Specific gravity, 60/60F: 0.989
  Weight, lbs./gallon @ 60F: 8.24
  Alkalinity as KOH, 1% solution (methyl orange indicator): 0.24
  pH, 10% solution: 10.0
  Free fatty acids, %: 3.0
  Surface tension, 1% solution: 28.5
  Appearance: clear, brown semi-gel
  Appearance, 10% solution in water: hazy, light brown liquid

EMULAMID AD-50 is a fatty diethanol amide that functions as a lubricant, emulsifier and rust inhibitor additive in water soluble cutting and grinding fluids and stamping, drawing and forging compounds. This is a biodegradable additive that offers unique solubility characteristics in both water and oil. This is a biodegradable additive that offers unique solubility characteristics in both water and oil.

Emulsifiable oils may be compounded with a reduced amount of emulsifier when EMULAMID AD-50 is added and still benefit from increased lubricity and a thicker, richer consistency. By adding a small amount of water to soluble oils containing EMULAMID AD-50, it is possible to form a stable thixotropic gel that will suspend insoluble solids such as powdered graphite.

EMULAMID TO-21 Lubricant-Emulsifier:

Typical Properties:
  Free Alkanolamine (as DEA): 25
  Acid Number: 15
  Free Fatty Acid (as Oleic): 7.5
  pH, 10% Solution: 10.0
  Specific Gravity 60/60F: 0.991
  Weight, Pounds/gallon at 60F: 8.25
  Appearance: Amber Liquid

EMULAMID TO-21 is a Tall Oil fatty amide designed for use in most water extendable metalworking coolant formulations. The addition of 5% to 20% EMULAMID TO-21 not only functions as a secondary emulsifier but also imparts corrosion protection, added lubricity and metal wetting to soluble oil, semi-synthetic and synthetic formulations.

When used in Synthetic and Semi-Synthetic coolants, EMULAMID TO-21 can serve as a primary or secondary emulsifier, while also improving boundary lubrication, wetting action, cooling capabilities and corrosion protection of the fluid.

MAYCO OIL & CHEMICAL CO.: MAYSOL Emulsifier Bases:

MAYSOL 410 Emulsifier Bases:

Typical Inspections:
  Specific gravity, 60/60F: 0.989
  Weight, lbs./gal. at 60F: 8.24
  Flash, COC, F: 350
  Fire, COC, F: 370
  pH(18% in naphthenic oil, mixed with water at 20:1 ratio):9.0
  Appearance and consistency: heavy, clear, brown oil

MAYSOL 410 Emulsifier Base is an economically priced emulsifier concentrate for the formulator who demands maximum emulsion stability at the lowest possible price.
MAYSOL 410 utilizes three distinctly different types of emulsifier in a synergistic combination that imparts excellent soluble oil mix stability in both hard and soft waters. Natural foam inhibitors allow for efficient wetting action with optimum foam control. Rust control is provided by a semi-polar rust inhibitor system that assures the protection of your customers' valuable machine tools.

MAYSOL 767 & 812:
For naphthenic oils (hydro-treated and/or solvent extracted) and paraffinic oils.

Typical Properties:
MAYSOL 767:
  Specific Gravity 60/60F: 1.006
  Wt. lbs./gallon @ 60F: 8.4
  Free Fatty Acid (as Oleic): 19-20
  Total Alkalinity (as KOH): 2.9

MAYSOL 812:
  Specific Gravity 60/60F: 1.006
  Wt. lbs./gallon @ 60F: 8.4
  Free Fatty Acid (as Oleic): 19-20
  Total Alkalinity (as KOH): 2.9

These two emulsifier bases were developed for use in the new hydro-treated and/or solvent extracted naphthenic base oils. Blends of 15% emulsifier base and 85% base oil will yield stable, non-skinning soluble oils that are not only stable under a wide range of temperatures (including freeze thaw tests), but also of any heavy metals, phenols, nitrites, etc. MAYSOL 767 and MAYSOL 812 can be used as emulsifier bases for paraffinic base oils if oleic acid or tall oil fatty acid (0.5 to 1.0%) is added to the base oil.

MAZER CHEMICALS: AVANEL S Surfactants:

Typical Properties of AVANEL S Surfactants:

<u>AVANEL S:</u>

30:
   Average Molecular Weight: 420
   % Solids: 34-36
   Appearance: White Paste
   Color, APHA: <100
   Specific Gravity @ 25C: 1.06
   Viscosity, cps @ 25C: 360
   pH: 6.8-7.6
   Flash Point, F: 130
   Solidification Point, C: 25

70:
   Average Molecular Weight: 600
   % Solids: 34-36
   Appearance: Clear Liquid
   Color, APHA: <100
   Specific Gravity @ 25C: 1.07
   Viscosity, cps @ 25C: 270
   pH: 6.8-7.6
   Flash Point, F: >200
   Solidification Point, C: -1

90:
   Average Molecular Weight: 690
   % Solids: 34-36
   Appearance: Clear Liquid
   Color, APHA: <100
   Specific Gravity @ 25C: 1.07
   Viscosity, cps @ 25C: 60
   pH: 6.8-7.6
   Flash Point, F: >200
   Solidification Point, C: -1

150:
   Average Molecular Weight: 950
   % Solids: 34-36
   Appearance: Clear Liquid
   Color, APHA: <100
   Specific Gravity @ 25C: 1.07
   Viscosity, cps @ 25C: 70
   pH: 6.8-7.6
   Flash Point, F: >200
   Solidification Point, C: -1

AVANEL S surfactants are presently available in varying polyoxyethylene chain lengths. Custom variations can also be made on special order. All of the products are approximately 35% solids and are readily biodegradable. The products are almost colorless, odorless, and are very mild to the skin.

MAZER CHEMICALS: AVANEL S Surfactants(Continued):

AVANEL S Surfactants represent a family of anionics, which may be generally described as sodium linear alkyl polyether sulfonates. Their unusual performance properties stem from their novel composition in which the sulfonate group is attached terminally to a polyoxyethylene chain of varying length.

Applications for AVANEL S Surfactants:
* Personal care formulations--conditioning shampoos, facial and germicidal scrubs, skin creams and lotions, oxidative skin and hair treatments, perfume solubilization, hair relaxers and curling solutions.

* Acid and alkaline cleaners and degreasers, metal pickling baths, surface preparation for electroplating, cleaners and rinses for printed circuit boards.

* USDA approved dairy and food cleaners, disinfectant cleaners, automatic dishwash detergents

* Enhanced oil recovery processes (including CO_2 and alkaline flooding), textile scouring, fiber processing, agricultural chemical emulsification, paper and pulp processing aids, water-based inks

* Emulsion polymerization of acrylic copolymer systems for use as coatings, polishes, sealants, textile finishes and adhesives

Key Features of AVANEL S Surfactants:
* Excellent hard water tolerance
* Hydrolytic stability over the entire pH range
* Complete biodegradability
* Oxidative stability in hypochlorite and oxygen bleaches
* Thermal stability
* High electrolyte tolerance
* Excellent rinsability
* Promotes sheeting action
* Extremely mild to skin
* Good emulsification characteristics
* Low critical micelle concentrations

MAZER CHEMICALS: MACOL Nonionic Surfactants:

The MACOL, nonionic-type surface active agents, show hydrophilic (water loving) properties through the presence of ether linkages that are capable of hydrogen bonding with water. The ether linkages are manufactured by polymerizing ethylene oxide on a hydrophobic (oil soluble) base. As the number of ether linkages increase on the hydrophobic base, the capability for hydrogen bonding with water increases, giving greater water solubility. The number of moles of ethylene oxide on the hydrophobic base is expressed numerically at the end of the product name.

The MACOL Nonionic Surfactants are excellent emulsifiers, detergents, and wetting agents. They are useful dispersants, solubilizers and coupling agents in cosmetic, textile, metalworking, household, industrial and various diversified areas.

<u>MACOL Fatty Alcohol Ethers:</u>

CA-2:
   Form @ 25C: Solid
   Melt Point C: 38
   Iodine Value: 0.5
   Hydroxyl Value: 170
   HLB Value: 4.9
   CTFA Name: Ceteth-2

CA-10:
   Form @ 25C: Solid
   Melt Point C: 41
   Iodine Value: 0.5
   Hydroxyl Value: 95
   HLB Value: 12.3
   CTFA Name: Ceteth-10

CSA-2:
   Form @ 25C: Solid
   Melt Point C: 39
   Iodine Value: 0.5
   Hydroxyl Value: 160
   HLB Value: 4.9
   CTFA Name: Ceteareth-2

CSA-4:
   Form @ 25C: Solid
   Melt Point C: 38
   Iodine Value: 0.5
   Hydroxyl Value: 128
   HLB Value: 7.9
   CTFA Name: Ceteareth-4

MAZER CHEMICALS: MACOL Nonionic Surfactants(Continued):

MACOL Fatty Alcohol Ethers:

CSA-10:
   Form @ 25C: Solid
   Melt Point C: 38
   Iodine Value: 0.5
   Hydroxyl Value: 80
   HLB Value: 12.3
   CTFA Name: Ceteareth-10

CSA-15:
   Form @ 25C: Solid
   Melt Point C: 38
   Iodine Value: 0.5
   Hydroxyl Value: 65
   HLB Value: 14.2
   CTFA Name: Ceteareth-15

CSA-20:
   Form @ 25C: Solid
   Melt Point C: 40
   Iodine Value: 0.5
   Hydroxyl Value: 52
   HLB Value: 15.2
   CTFA Name: Ceteareth-20

CSA-40:
   Form @ 25C: Solid
   Melt Point C: 40
   Iodine Value: 0.5
   Hydroxyl Value: 30
   HLB Value: 16.8
   CTFA Name: Ceteareth-40

LA-4:
   Form @ 25C: Liquid
   Melt Point C: 12
   Iodine Value: 0.1
   Hydroxyl Value: 155
   HLB Value: 9.5
   CTFA Name: Laureth-4

MAZER CHEMICALS: MACOL Nonionic Surfactants(Continued):

MACOL Fatty Alcohol Ethers:

LA-790*:
    Form @ 25C: Liquid
    Melt Point C: 5
    Iodine Value: 0.1
    HLB Value: 10.8
    CTFA Name: Laureth-7
    * 90% active in water

LA-9:
    Form @ 25C: Paste
    Melt Point C: 26
    Iodine Value: 0.1
    Hydroxyl Value: 95
    HLB Value: 13.3
    CTFA Name: Laureth-9

LA-12:
    Form @ 25C: Solid
    Melt Point C: 30
    Iodine Value: 0.1
    Hydroxyl Value: 75
    HLB Value: 14.6
    CTFA Name: Laureth-12

LA-23:
    Form @ 25C: Solid
    Melt Point C: 40
    Iodine Value: 0.1
    Hydroxyl Value: 47
    HLB Value: 16.4
    CTFA Name: Laureth-23

OA-2:
    Form @ 25C: Liquid
    Melt Point C: <0
    Iodine Value: 70
    Hydroxyl Value: 170
    HLB Value: 3.8
    CTFA Name: Oleth-2

OA-4:
    Form @ 25C: Liquid
    Melt Point C: <0
    Iodine Value: 53
    Hydroxyl Value: 128
    HLB Value: 8.0
    CTFA Name: Oleth-4

MAZER CHEMICALS: MACOL Nonionic Surfactants(Continued):

MACOL Fatty Alcohol Ethers:

OA-5:
   Form @ 25C: Liquid
   Melt Point C: 5
   Iodine Value: 53
   Hydroxyl Value: 125
   HLB Value: 8.2
   CTFA Name: Oleth-5

OA-10:
   Form @ 25C: Liquid
   Melt Point C: 16
   Iodine Value: 33
   Hydroxyl Value: 80
   HLB Value: 12.5
   CTFA Name: Oleth-10

OA-20:
   Form @ 25C: Solid
   Melt Point C: 30
   Iodine Value: 23
   Hydroxyl Value: 58
   HLB Value: 14.7
   CTFA Name: Oleth-20

SA-2:
   Form @ 25C: Solid
   Melt Point C: 43
   Iodine Value: 0.1
   Hydroxyl Value: 158
   HLB Value: 4.7
   CTFA Name: Steareth-2

SA-5:
   Form @ 25C: Solid
   Melt Point C: 41
   Iodine Value: 0.1
   Hydroxyl Value: 116
   HLB Value: 9.0
   CTFA Name: Steareth-5

SA-10:
   Form @ 25C: Solid
   Melt Point C: 40
   Iodine Value: 0.1
   Hydroxyl Value: 80
   HLB Value: 12.3
   CTFA Name: Steareth-10

MAZER CHEMICALS: MACOL Nonionic Surfactants(Continued):

MACOL Fatty Alcohol Ethers:

SA-15:
   Form @ 25C: Solid
   Melt Point C: 38
   Iodine Value: 0.1
   Hydroxyl Value: 64
   HLB Value: 14.3
   CTFA Name: Steareth-15

SA-20:
   Form @ 25C: Solid
   Melt Point C: 39
   Iodine Value: 0.1
   Hydroxyl Value: 52
   HLB Value: 15.4
   CTFA Name: Steareth-20

SA-40:
   Form @ 25C: Solid
   Melt Point C: 40
   Iodine Value: 0.1
   Hydroxyl Value: 32
   HLB Value: 17.4
   CTFA Name: Steareth-40

MACOL Nonionic Surfactants:

DNP-5:
   Form @ 25C: Liquid
   Viscosity @ 25C in cps: 385
   Pour Point C: -10
   HLB Value: 8.2

DNP-10:
   Form @ 25C: Liquid
   Viscosity @ 25C in cps: 390
   Pour Point C: 0
   HLB Value: 11.3

DNP-15:
   Form @ 25C: Paste
   Pour Point C: 30
   HLB Value: 13.0

DNP-150:
   Form @ 25C: Flake
   Pour Point C: 55
   HLB Value: 19.0

MAZER CHEMICALS: MACOL Nonionic Surfactants(Continued):

MACOL:

NP-4:
   Form @ 25C: Liquid
   Viscosity @ 25C in cps: 350
   Pour Point C: -27
   HLB Value: 8.9

NP-5:
   Form @ 25C: Liquid
   Viscosity @ 25C in cps: 320
   Pour Point C: -27
   HLB Value: 10.0

NP-6:
   Form @ 25C: Liquid
   Viscosity @ 25C in cps: 300
   Pour Point C: -28
   HLB Value: 10.9

NP-8:
   Form @ 25C: Liquid
   Viscosity @ 25C in cps: 260
   Pour Point C: 5
   HLB Value: 12.3

NP-9.5:
   Form @ 25C: Liquid
   Viscosity @ 25C in cps: 275
   Pour Point C: 5
   HLB Value: 12.9

NP-11:
   Form @ 25C: Liquid
   Viscosity @ 25C in cps: 275
   Pour Point C: 14
   HLB Value: 13.7

NP-12:
   Form @ 25C: Liquid
   Viscosity @ 25C in cps: 325
   Pour Point C: 17
   HLB Value: 14.0

NP-15:
   Form @ 25C: Paste
   Pour Point C: 26
   HLB Value: 15.0

MAZER CHEMICALS: MACOL Nonionic Surfactants(Continued):

MACOL:

NP-20:
    Form @ 25C: Solid
    Pour Point C: 30
    HLB Value: 16.0

NP-20(70):
    Form @ 25C: Liquid
    Pour Point C: 0
    HLB Value: 16.0

NP-30(70):
    Form @ 25C: Liquid
    Pour Point C: 4
    HLB Value: 17.2

NP-100:
    Form @ 25C: Solid
    Pour Point C: 54
    HLB Value: 19.0

OP-3:
    Form @ 25C: Liquid
    Pour Point C: -23
    HLB Value: 7.8

OP-5:
    Form @ 25C: Liquid
    Pour Point C: -26
    HLB Value: 10.4

OP-8:
    Form @ 25C: Liquid
    Pour Point C: -5
    HLB Value: 12.3

OP-10:
    Form @ 25C: Liquid
    Pour Point C: 8
    HLB Value: 13.4

OP-12:
    Form @ 25C: Liquid
    Pour Point C: 16
    HLB Value: 14.6

OP-16(75):
    Form @ 25C: Liquid
    Pour Point C: 13
    HLB Value: 15.8

MACOL CHEMICALS: MACOL Nonionic Surfactants(Continued):

MACOL:

OP-30(70):
   Form @ 25C: Liquid
   Pour Point C: 2
   HLB Value: 17.3

OP-40(70):
   Form @ 25C: Liquid
   Pour Point C: -4
   HLB Value: 17.9

TD-3:
   Form @ 25C: Liquid
   Pour Point C: -32
   HLB Value: 8.0

TD-8:
   Form @ 25C: Liquid
   Pour Point C: 8
   HLB Value: 12.4

TD-10:
   Form @ 25C: Liquid
   Pour Point C: 10
   HLB Value: 13.6

TD-12:
   Form @ 25C: Liquid
   Pour Point C: 14
   HLB Value: 14.1

TD-100:
   Form @ 25C: Solid
   Pour Point C: 55
   HLB Value: 18.9

TD-610:
   Form @ 25C: Liquid
   Pour Point C: 6
   HLB Value: 11.3

MAZER CHEMICALS: MACOL and MAZAWAX Cosmetic Emulsifiers:

MACOL and MAZAWAX Nonionic Cosmetic Emulsifiers were developed particularly for creams, lotions, and ointments for the Cosmetic and Pharmaceutical Industry. Their unique emulsifying characteristics produce emulsions of superior stability and esthetic quality under a wide range of acidic and alkaline conditions.

Product:

MACOL 123:
    Form @ 25C: Flake
    Melt Point C: 52
    Saponification Value: 2
    CTFA: Cetearyl Alcohol and Ceteareth-20 and Ceteareth-10

MACOL 124:
    Form @ 25C: Flake
    Melt Point C: 52
    Saponification Value: 2
    CTFA: Cetearyl Alcohol and Ceteareth-20

MACOL 125:
    Form @ 25C: Flake
    Melt Point C: 60
    Saponification Value: 2
    CTFA: Stearyl Alcohol and Ceteareth-20

MACOL CPS:
    Form @ 25C: Flake
    Melt Point C: 51
    Saponification Value: 12
    CTFA: Cetearyl Alcohol and Polysorbate 60 and PEG 150 Stearate and Ceteareth-20

MAZAWAX 163R:
    Form @ 25C: Flake
    Melt Point C: 51
    Saponification Value: 12
    CTFA: Cetearyl Alcohol and Polysorbate 60

MAZER CHEMICALS: MAFO Amphoteric Surfactants:

MAFO Amphoteric Surfactants act as an anion or cation depending on the pH of the system. This makes them compatible with either strong alkali or acid systems and with commonly used surfactants. MAFO Surfactants are biodegradable, tolerate hard water and have a chelating action which clarifies many systems. They act as solubilizers for difficult materials, such as phenolic germicides and phosphated alcohols and tolerate high levels of inorganic salts. MAFO Surfactants provide mild substantive properties under essentially neutral conditions which result in emolliency, or softening, of the skin.

MAFO:

13:
   Percent Solids: 70
   Diluent: water
   Description: Potassium Salt of a Complex Amine Carboxylate

13MOD1:
   Percent Solids: 90
   Diluent: water
   Description: Potassium Salt of a Complex Amine Carboxylate

C:
   Percent Solids: 35
   Diluent: water
   Description: Cocamidopropyl Betaine

CAB:
   Percent Solids: 35
   Diluent: water
   Description: Cocamidopropyl Betaine

CAB SP:
   Percent Solids: 43
   Duluent: water
   Description: Cocamidopropyl Betaine

CAB 425:
   Percent Solids: 42.5
   Diluent: water
   Description: Cocamidopropyl Betaine

CB 40:
   Percent Solids: 40
   Diluent: water
   Description: Coco Betaine

CFA 35:
   Percent Solids: 35
   Diluent: water
   Description: Cocamidopropyl Betaine

MAZER CHEMICALS: MAFO Amphoteric Surfactants(Continued):

MAFO:

CSB:
   Percent Solids: 35
   Diluent: water
   Description: Cocamidopropyl Hydroxysultaine

CSB 50:
   Percent Solids: 50
   Diluent: water
   Description: Cocamidopropyl Hydroxysultaine

CSB W:
   Percent Solids: 50
   Diluent: water
   Description: Cocamidopropyl Hydroxysultaine

KCOSB 50:
   Percent Solids: 50
   Diluent: water
   Description: Cocamidopropyl Hydroxysultaine

LMAB:
   Percent Solids: 35
   Diluent: water
   Description: Lauramidopropyl Betaine

OB:
   Percent Solids: 50
   Diluent: water
   Description: Oleyl Betaine

SBAO 110:
   Percent Solids: 42
   Diluent: water
   Description: Lime Dispersant

MAZER CHEMICALS: MAPEG Polyethylene Glycol Esters:

MAPEG Polyethylene Glycol Esters are mono and diesters of various fatty acids. They offer graduated hydrophilic to lipophilic surface active properties, which make them useful as primary and secondary nonionic surfactants, with stability over a wide range of formulating conditions. The MAPEG esters in the range of 200 to 1540 molecular weight are the most versatile in regard to emulsification properties.

In addition to being effective emulsifiers, the MAPEG esters are highly emollient and are excellent solubilizers in bath oils and fragrance compositions. They provide washability in anhydrous formulations such as hair preparations and ointments. The higher molecular weight MAPEG distearates are outstanding thickening agents in aqueous systems.

MAPEG:

EGMS:
   Description: Ethylene Glycol Monostearate
   Form @ 25C: Flake
   Pour Point C: 56
   Saponification Value: 184
   Maximum Acid Value: 4
   HLB Value: 2.9

EGDS:
   Description: Ethylene Glycol Distearate
   Form @ 25C: Flake
   Pour Point C: 63
   Saponification Value: 195
   Maximum Acid Value: 6
   HLB Value: 1.4

200 MS:
   Description: PEG 200 Monostearate
   Form @ 25C: Solid
   Pour Point C: 33
   Saponification Value: 125
   Maximum Acid Value: 5
   HLB Value: 8.0

200 DS:
   Description: PEG 200 Distearate
   Form @ 25C: Solid
   Pour Point C: 34
   Saponification Value: 160
   Maximum Acid Value: 10
   HLB Value: 11.5

MAZER CHEMICALS: MAPEG Polyethylene Glycol Esters(Continued):

MAPEG:

400 MS:
    Description: PEG 400 Monostearate
    Form @ 25C: Solid
    Pour Point C: 33
    Saponification Value: 88
    Maximum Acid Value: 5
    HLB Value: 11.5

400 DS:
    Description: PEG 400 Distearate
    Form @ 25C: Solid
    Pour Point C: 36
    Saponification Value: 124
    Maximum Acid Value: 10
    HLB Value: 8.1

600 MS:
    Description: PEG 600 Monostearate
    Form @ 25C: Solid
    Pour Point C: 36
    Saponification Value: 66
    Maximum Acid Value: 5
    HLB Value: 13.6

600 DS:
    Description: PEG 600 Distearate
    Form @ 25C: Solid
    Pour Point C: 41
    Saponification Value: 98
    Maximum Acid Value: 10
    HLB Value: 10.6

1000 MS:
    Description: PEG 1000 Monostearate
    Form @ 25C: Solid
    Pour Point C: 42
    Saponification Value: 45
    Maximum Acid Value: 5
    HLB Value: 15.7

1500 MS:
    Description: PEG 1500 Monostearate
    Form @ 25C: Solid
    Pour Point C: 37
    Saponification Value: 62
    Maximum Acid Value: 5
    HLB Value: 16.1

MAZER CHEMICALS: MAPEG Polyethylene Glycol Esters(Continued):

MAPEG:

1540 DS:
  Description: PEG 1540 Distearate
  Form @ 25C: Flake
  Pour Point C: 45
  Saponification Value: 53
  Maximum Acid Value: 10
  HLB Value: 14.8

S-40:
  Description: PEG 1760 Monostearate
  Form @ 25C: Flake
  Pour Point C: 44
  Saponification Value: 30
  Maximum Acid Value: 1
  HLB Value: 17.2

S-100:
  Description: PEG 4400 Monostearate
  Form @ 25C: Flake
  Pour Point C: 50
  Saponification Value: 16
  Maximum Acid Value: 1
  HLB Value: 18.7

S-150:
  Description: PEG 6000 Monostearate
  Form @ 25C: Flake
  Pour Point C: 51
  Saponification Value: 9.5
  Maximum Acid Value: 1
  HLB Value: 19.0

6000 DS:
  Description: PEG 6000 Distearate
  Form @ 25C: Flake
  Pour Point C: 56
  Saponification Value: 20
  Maximum Acid Value: 10
  HLB Value: 18.4

200 ML:
  Description: PEG 200 Monolaurate
  Form @ 25C: Liquid
  Pour Point C: 5
  Saponification Value: 148
  Maximum Acid Value: 5
  HLB Value: 9.3

MAZER CHEMICALS: MAPEG Polyethylene Glycol Esters(Continued):

MAPEG:

200 DL:
  Description: PEG 200 Dilaurate
  Form @ 25C: Liquid
  Pour Point C: 10
  Saponification Value: 185
  Maximum Acid Value: 10
  HLB Value: 7.6

400 ML:
  Description: PEG 400 Monolaurate
  Form @ 25C: Liquid
  Pour Point C: 12
  Saponification Value: 93
  Maximum Acid Value: 5
  HLB Value: 13.2

400 DL:
  Description: PEG 400 Dilaurate
  Form @ 25C: Liquid
  Pour Point C: 18
  Saponification Value: 135
  Maximum Acid Value: 10
  HLB Value: 10.8

600 ML:
  Description: PEG 600 Monolaurate
  Form @ 25C: Paste
  Pour Point C: 23
  Saponification Value: 70
  Maximum Acid Value: 5
  HLB Value: 14.8

600 DL:
  Description: PEG 600 Dilaurate
  Form @ 25C: Paste
  Pour Point C: 24
  Saponification Value: 105
  Maximum Acid Value: 10
  HLB Value: 12.2

200 MO:
  Description: PEG 200 Monooleate
  Form @ 25C: Liquid
  Pour Point C: -26
  Saponification Value: 120
  Maximum Acid Value: 5
  HLB Value: 8.3

MAZER CHEMICALS: MAPEG Polyethylene Glycol Esters(Continued):

MAPEG:

200 DO:
   Description: PEG 200 Dioleate
   Form @ 25C: Liquid
   Pour Point C: 2
   Saponification Value: 150
   Maximum Acid Value: 10
   HLB Value: 6.0

400 MO:
   Description: PEG 400 Monooleate
   Form @ 25C: Liquid
   Pour Point C: 5
   Saponification Value: 84
   Maximum Acid Value: 5
   HLB Value: 11.8

400 DO:
   Description: PEG 400 Dioleate
   Form @ 25C: Liquid
   Pour Point C: 6
   Saponification Value: 118
   Maximum Acid Value: 10
   HLB Value: 8.8

600 MO:
   Description: PEG 600 Monooleate
   Form @ 25C: Liquid
   Pour Point C: 18
   Saponification Value: 65
   Maximum Acid Value: 5
   HLB Value: 13.6

600 DO:
   Description: PEG 600 Dioleate
   Form @ 25C: Liquid
   Pour Point C: 19
   Saponification Value: 98
   Maximum Acid Value: 10
   HLB Value: 10.3

200 MOT:
   Description: PEG 200 Monotallate
   Form @ 25C: Liquid
   Pour Point C: -22
   Saponification Value: 120
   Maximum Acid Value: 5
   HLB Value: 8.3

MAZER CHEMICALS: MAPEG Polyethylene Glycol Esters(Continued):

MAPEG:

200 DOT:
  Description: PEG 200 Ditallate
  Form @ 25C: Liquid
  Pour Point C: -18
  Saponification Value: 150
  Maximum Acid Value: 10
  HLB Value: 6.0

400 MOT:
  Description: PEG 400 Monotallate
  Form @ 25C: Liquid
  Pour Point C: 5
  Saponification Value: 84
  Maximum Acid Value: 5
  HLB Value: 11.8

400 DOT:
  Description: PEG 400 Ditallate
  Form @ 25C: Liquid
  Pour Point C: 6
  Saponification Value: 118
  Maximum Acid Value: 10
  HLB Value: 8.8

600 DOT:
  Description: PEG 600 Ditallate
  Form @ 25C: Liquid
  Pour Point C: 19
  Saponification Value: 98
  Maximum Acid Value: 10
  HLB Value: 10.3

TAO-15:
  Description: PEG 660 Monotallate
  Form @ 25C: Liquid
  Pour Point C: 17
  Saponification Value: 60
  Maximum Acid Value: 5
  HLB Value: 13.8

CO-16H:
  Description: PEG 700 Hydrogenated Castor Oil
  Form @ 25C: Liquid
  Pour Point C: 7
  Saponification Value: 105
  Maximum Acid Value: 2
  HLB Value: 8.6

MAZER CHEMICALS: MAPEG Polyethylene Glycol Esters(Continued):

MAPEG:

CO-25:
   Description: PEG 1100 Castor Oil
   Form @ 25C: Liquid
   Pour Point C: 5
   Saponification Value: 83
   Maximum Acid Value: 2
   HLB Value: 10.8

CO-25H:
   Description: PEG 1100 Hydrogenated Castor Oil
   Form @ 25C: Liquid
   Pour Point C: 5
   Saponification Value: 82
   Maximum Acid Value: 2
   HLB Value: 10.8

CO-30:
   Description: PEG 1320 Castor Oil
   Form @ 25C: Liquid
   Pour Point C: 9
   Saponification Value: 75
   Maximum Acid Value: 2
   HLB Value: 11.8

CO-36:
   Description: PEG 1580 Castor Oil
   Form @ 25C: Liquid
   Pour Point C: 12
   Saponification Value: 73
   Maximum Acid Value: 2
   HLB Value: 12.6

CO-200:
   Description: PEG 8800 Castor Oil
   Form @ 25C: Solid
   Pour Point C: 50
   Saponification Value: 17.5
   Maximum Acid Value: 2
   HLB Value: 18.1

MAZER CHEMICALS: MAZAMIDE Alkanolamides:

MAZAMIDES are alkanolamides introduced and manufactured by Mazer Chemicals for the chemical specialties and cosmetics industries. Their basic uses are for thickening, emulsification, foam boosting and rust inhibition, for a wide range of products including: liquid detergents, shampoos and industrial lubricants.

MAZAMIDE:

65:
- Form @ 25C: Liquid
- Type: 2:1
- Fatty Acid: Mixed
- % FFA: 25
- % Free Amine: 21

68:
- Form @ 25C: Liquid
- Type: 2:1
- Fatty Acid: Coconut
- % FFA: 29
- % Free Amine: 27

70:
- Form @ 25C: Liquid
- Type: 2:1
- Fatty Acid: Coconut
- % FFA: 3
- % Free Amine: 30

80:
- Form @ 25C: Liquid
- Type: 1:1
- Fatty Acid: Coconut
- % FFA: 0.5
- % Free Amine: 6

124:
- Form @ 25C: Liquid
- Type: 1:1
- Fatty Acid: Mixed
- % FFA: 0.5
- % Free Amine: 7.5

524:
- Form @ 25C: Liquid
- Type: 2:1
- Fatty Acid: Mixed
- % FFA: 27
- % Free Amine: 23

MAZER CHEMICALS: MAZAMIDE Alkanolamides(Continued):

MAZAMIDE:

1214:
   Form @ 25C: Liquid
   Type: 2:1
   Fatty Acid: Lauric
   % FFA: 5
   % Free Amine: 23

1281:
   Form @ 25C: Liquid
   Type: 2:1
   Fatty Acid: Mixed
   % FFA: 53
   % Free Amine: 22

C2:
   Form @ 25C: Liquid
   Type: POE3
   Fatty Acid: Coconut
   % FFA: 0.5
   % Free Amine: 0

C5:
   Form @ 25C: Liquid
   Type: POE6
   Fatty Acid: Coconut
   % FFA: 0.5
   % Free Amine: 0

CS148:
   Form @ 25C: Liquid
   Type: 1:1
   Fatty Acid: Coconut
   % FFA: 0.5
   % Free Amine: 7

CCO:
   Form @ 25C: Liquid
   Type: 2:1
   Fatty Acid: Coconut
   % FFA: 5
   % Free Amine: 28

CFAM:
   Form @ 25C: Flake
   Type: 1:1
   Fatty Acid: Coconut
   % FFA: 0.5
   % Free Amine: 2

MAZER CHEMICALS: MAZAMIDE Alkanolamides(Continued):

MAZAMIDE:

CMEA:
    Form @ 25C: Flake
    Type: 1:1
    Fatty Acid: Coconut
    % FFA: 0.5
    % Free Amine: 2

CMEA Extra:
    Form @ 25C: Flake
    Type: 1:1
    Fatty Acid: Coconut
    % FFA: 0.5
    % Free Amine: 2

J10:
    Form @ 25C: Liquid
    Type: 2:1
    Fatty Acid: Mixed
    % FFA: 73
    % Fatty Amine: 19

JR100:
    Form @ 25C: Liquid
    Type: 2:1
    Fatty Acid: Mixed
    % FFA: 48
    % Fatty Amine: 20

JR300:
    Form @ 25C: Liquid
    Type: 2:1
    Fatty Acid: Mixed
    % FFA: 70
    % Fatty Amine: 20

JR400:
    Form @ 25C: Liquid
    Type: 2:1
    Fatty Acid: Mixed
    % FFA: 55
    % Free Amine: 20

JT128:
    Form @ 25C: Liquid
    Type: 2:1
    Fatty Acid: Coconut
    % FFA: 1
    % Free Amine: 7

MAZER CHEMICALS: MAZAMIDE Alkanolamides(Continued):

MAZAMIDE:

L5:
   Form @ 25C: Solid
   Type: POE6
   Fatty Acid: Lauric
   % FFA: 0.5
   % Free Amine: 0

L298:
   Form @ 25C: Liquid
   Type: 2:1
   Fatty Acid: Lauric
   % FFA: 5
   % Free Amine: 35

LLD:
   Form @ 25C: Liquid
   Type: 2:1
   Fatty Acid: Linoleic
   % FFA: 1
   % Free Amine: 6

LM:
   Form @ 25C: Liquid
   Type: 2:1
   Fatty Acid: Lauric
   % FFA: 7
   % Free Amine: 31

LM20:
   Form @ 25C: Liquid
   Type: 2:1
   Fatty Acid: Coconut
   % FFA: 10
   % Free Amine: 27

LS196:
   Form @ 25C: Liquid
   Type: 1:1
   Fatty Acid: Lauric
   % FFA: 0.5
   % Free Amine: 7

O20:
   Form @ 25C: Liquid
   Type: 2:1
   Fatty Acid: Oleic
   % FFA: 7
   % Free Amine: 24

MAZER CHEMICALS: MAZAMIDE Alkanolamides(Continued):

MAZAMIDE:

PCS:
    Form @ 25C: Liquid
    Type: 2:1
    Fatty Acid: Mixed
    % FFA: 62
    % Free Amine: 12

RO:
    Form @ 25C: Liquid
    Type: 2:1
    Fatty Acid: Mixed
    % FFA: 80
    % Free Amine: 34

T20:
    Form @ 25C: Liquid
    Type: 2:1
    Fatty Acid: Tall Oil
    % FFA: 5
    % Free Amine: 21

TC:
    Form @ 25C: Liquid
    Type: 2:1
    Fatty Acid: Lauric
    % FFA: 19
    % Free Amine: 30

SCD:
    Form @ 25C: Liquid
    Type: 1:1
    Fatty Acid: Mixed
    % FFA: 0.5
    % Free Amine: 7

SS10:
    Form @ 25C: Liquid
    Type: 1:1
    Fatty Acid: Linoleic
    % FFA: 1.5
    % Free Amine: 6.5

SS20:
    Form @ 25C: Liquid
    Type: 2:1
    Fatty Acid: Linoleic
    % FFA: 1.5
    % Free Amine: 22

MAZER CHEMICALS: MAZAMIDE Alkanolamides(Continued):

MAZAMIDE:

SMEA:
    Form @ 25C: Flake
    Type: 1:1
    Fatty Acid: Stearic
    % FFA: 0.5
    % Free Amine: 2

WC Concentrate:
    Form @ 25C: Liquid
    Type: 1:1
    Fatty Acid: Coconut
    % FFA: 0.5
    % Free Amine: 6

MAZER CHEMICALS: MAZAWET Wetting Agents:

The MAZAWET Surfactants are commonly called "wetting agents" because of their ability to lower the surface tension between water and organic liquids.

MAZAWET:

DF:
    Form @ 25C: Liquid
    Surface Tension Dynes/Cm: 27.3
    Percent Activity: 100
    Ionic Nature: Nonionic

DOSS 70:
    Form @ 25C: Liquid
    Surface Tension Dynes/Cm: 29.0
    Percent Activity: 70
    Ionic Nature: Anionic

36:
    Form @ 25C: Liquid
    Surface Tension Dynes/Cm: 32.0
    Percent Activity: 92
    Ionic Nature: Nonionic

77:
    Form @ 25C: Liquid
    Surface Tension Dynes/Cm: 28.1
    Percent Activity: 100
    Ionic Nature: Nonionic

MAZER CHEMICALS: MAZEEN Ethoxylated Amines:

MAZEEN Surfactants are a group of tertiary amines substituted with two or more polyoxyethylene groups attached to the nitrogen. These products are essentially cationic in nature; however, with increasing ethylene oxide content, these products become nonionic in character.

MAZEEN Surfactants have applications as insecticide and herbicide emulsifiers, antistatic agents, rewetting agents, grease additives, textile lubricants, and as emulsifiers in lubricants, water based inks and cosmetic formulations.

MAZEEN:

C-2:
  Form @ 25C: Liquid
  Average Equivalent Weight: 285
  Color Gardner: 10
  Specific Gravity @ 25C: 0.874
  Surface Tension Dynes/Cm: 28

C-5:
  Form @ 25C: Liquid
  Average Equivalent Weight: 425
  Color Gardner: 10
  Specific Gravity @ 25C: 0.976
  Surface Tension Dynes/Cm: 33

C-10:
  Form @ 25C: Liquid
  Average Equivalent Weight: 645
  Color Gardner: 11
  Specific Gravity @ 25C: 1.017
  Surface Tension Dynes/Cm: 39

C-15:
  Form @ 25C: Liquid
  Average Equivalent Weight: 860
  Color Gardner @ 25C: 9
  Specific Gravity @ 25C: 1.042
  Surface Tension Dynes/Cm: 41

DBA-1:
  Form @ 25C: Liquid
  Average Equivalent Weight: 172
  Color Gardner @ 25C: 2
  Specific Gravity @ 25C: 0.860
  Surface Tension Dynes/Cm: 41

MAZER CHEMICALS: MAZEEN Ethoxylated Amines(Continued):

MAZEEN:

S-2:
   Form @ 25C: Liquid
   Average Equivalent Weight: 350
   Color Gardner: 14
   Specific Gravity @ 25C: 0.911
   Surface Tension Dynes/Cm: 26

S-5:
   Form @ 25C: Liquid
   Average Equivalent Weight: 500
   Color Gardner: 11
   Specific Gravity @ 25C: 0.951
   Surface Tension Dynes/Cm: 33

S-10:
   Form @ 25C: Liquid
   Average Equivalent Weight: 710
   Color Gardner: 11
   Specific Gravity @ 25C: 1.020
   Surface Tension Dynes/Cm: 40

S-15:
   Form @ 25C: Liquid
   Average Equivalent Weight: 930
   Color Gardner: 11
   Specific Gravity @ 25C: 1.040
   Surface Tension Dynes/Cm: 43

T-2:
   Form @ 25C: Liquid
   Average Equivalent Weight: 350
   Color Gardner: 7
   Specific Gravity @ 25C: 0.916
   Surface Tension Dynes/Cm: 29

T-5:
   Form @ 25C: Liquid
   Average Equivalent Weight: 500
   Color Gardner: 14
   Specific Gravity @ 25C: 0.966
   Surface Tension Dynes/Cm: 34

T-15:
   Form @ 25C: Liquid
   Average Equivalent Weight: 925
   Color Gardner @ 25C: 10
   Specific Gravity @ 25C: 1.028
   Surface Tension Dynes/Cm: 41

MAZER CHEMICALS: MAZOL Glycerol and Polyglycerol Esters:

MAZOL Glycerol and Polyglyceryl Esters have been developed to cover the widest range of oil and water emulsification. These esters are excellent emulsifiers for dietitic, bakery and confectionery food products, cosmetics, toiletries, pharmaceuticals, lubricants, mold release compounds and in plasticizers for synthetic fabrics and plastics. MAZOL Glycerol and Polyglycerol Esters are non-toxic and are easily metabolized to glycerol and fatty acids by human and animal ingestion. Many of these MAZOL Glycerol and Polyglycerol Esters are generally recognized as safe (GRAS) by the FDA for many direct food applications. The usage of these esters is approved and regulated under Title 21, CFR 172.854.

All food grade MAZOL Glycerol and Polyglycerol Esters are manufactured under the guidance and supervision of the Union of Orthodox Jewish Congregations in America and are available with Ⓤ Certification. They are designated with a "K" in the product name.

MAZOL:

80 MG-K:
Description: Ethoxylated Mono/Diglycerides
Form @ 25C: Amber Liquid
Saponification Value: 65-75
Maximum Acid Value: 2
HLB Value: 13.5

159:
Description: PEG 7 Glycerol Cocoate
Form @ 25C: Amber Liquid
Saponification Value: 82-98
Maximum Acid Value: 5
HLB Value: 13.0

165 C:
Description: Glycerol Monostearate Acid Stable
Form @ 25C: Tan Flake
Saponification Value: 90-100
Mono Ester Content: 20
Percent Free Glycerine: 5
Maximum Acid Value: 2
HLB Value: 11.2

MAZER CHEMICALS: MAZOL Glycerol and Polyglycerol Esters (Continued):

MAZOL:

300 K:
   Description: Glycerol Monooleate
   Form @ 25C: Amber Liquid
   Mono Ester Content: 40
   Maximum Acid Value: 2
   HLB Value: 3.8
1400:
   Description: Caprylic Capric Triglyceride
   Form @ 25C: Clear Liquid
   Saponification Value: 335-360
   Maximum Acid Value: 0.5
   HLB Value: 1.0
GMO:
   Description: Glycerol Monooleate
   Form @ 25C: Dark Liquid
   Mono Ester Content: 40
   Percent Free Glycerine: 5
   Maximum Acid Value: 2
   HLB Value: 3.8
GMO-K:
   Description: Glycerol Monooleate
   Form @ 25C: Tan Paste
   Mono Ester Content: 40
   Percent Free Glycerine: 2
   Maximum Acid Value: 2
   HLB Value: 3.8
GMO-#1:
   Description: Glycerol Monooleate
   Form @ 25C: Dark Liquid
   Mono Ester Content: 32
   Percent Free Glycerine: 5
   Maximum Acid Value: 3
   HLB Value: 3.8
GMR:
   Description: Glycerol Monoricinoleate
   Form @ 25C: Dark Liquid
   Saponification Value: 138-145
   Percent Free Glycerine: 7
   Maximum Acid Value: 7
   HLB Value: 6.0
GMS:
   Description: Glycerol Monoostearate
   Form @ 25C: Tan Flake
   Mono Ester Content: 34
   Percent Free Glycerine: 3
   Maximum Acid Value: 5
   HLB Value: 3.9

MAZER CHEMICALS: MAZON Ethoxylated Sorbitol Esters:

MAZON Ethoxylated Sorbitol Esters are hydrophilic emulsifiers and coupling agents which are frequently used as co-emulsifiers with the S-MAZ Sorbitan Fatty Acid Esters. They are dispersible in water and soluble in mineral oils, low molecular weight, aromatic solvents, alcohols and ketones. These emulsifiers find uses in agricultural pesticide and herbicide formulations, emulsion polymerization, metalworking lubricants, and die-cast lubricants.

MAZON:

1045A:

    Form @ 25C: Liquid
    Viscosity @ 25C in Cps: 260
    Saponification Value: 90
    HLB Value: 13.0
    Flashpoint PMCC F: >350

1086:

    Form @ 25C: Liquid
    Viscosity @ 25C in Cps: 200
    Saponification Value: 97
    HLB Value: 10.4
    Flashpoint PMCC F: >350

1096:

    Form @ 25C: Liquid
    Viscosity @ 25C in Cps: 240
    Saponification Value: 85
    HLB Value: 11.2
    Flashpoint PMCC F: >350

MAZER CHEMICALS: MAZOX Amine Oxides:

MAZOX Amine Oxides are nonionic surfactants at neutral or alkaline pH ranges. In acidic solutions, amine oxides are mildly cationic. They are also fully biodegradable and compatible with all types of surfactants, such as anionics, cationics, nonionics and amphoterics.

The unique surface active characteristics of conditioning, emolliency, emulsification, foam boosting and viscosity building of the MAZOX Amine Oxides make these compounds the prime choice in applications ranging from cosmetics and toiletries to household and industrial uses.

MAZOX Amine Oxides can be used at low levels which relate to excellent cost savings and high performance.

Trade Name:

MAZOX CAPA:
    Form @ 25C: Liquid
    Percent: Amine Oxide: 30
    Peroxide Value: 100 max.

MAZOX CDA:
    Form @ 25C: Liquid
    Percent: Amine Oxide: 30
    Peroxide Value: 100 max.

MAZOX KCAO:
    Form @ 25C: Liquid
    Percent: Amine Oxide: 33
    Peroxide Value: 100 max.

MAZOX LDA:
    Form @ 25C: Liquid
    Percent: Amine Oxide: 30
    Peroxide Value: 100 max.

MAZOX MDA:
    Form @ 25C: Liquid
    Percent: Amine Oxide: 30
    Peroxide Value: 100 max.

MAZOX ODA:
    Form @ 25C: Liquid
    Percent: Amine Oxide: 50
    Peroxide Value: 150 max.

MAZOX SDA:
    Form @ 25C: Paste
    Percent: Amine Oxide: 25
    Peroxide Value: 100 max.

MAZER CHEMICALS: S-MAZ Sorbitan Fatty Acid Esters:

The S-MAZ Sorbitan Fatty Acid Esters are nonionic, lipophilic surfactants used for preparing excellent water in oil emulsions. They find use as antistats, textile softeners, lubricants, process defoamers, opacifiers, and co-emulsifiers. Together with the T-MAZ series of surfactants, they are very desirable as oil in water emulsifers for cosmetics, food formulations, industrial oils and household products. Some S-MAZ Sorbitan Fatty Acid Esters are also available for food use with a Kosher Ⓤ Certification. They are designated with a "K" in the product name.

S-MAZ:

20:
   Sorbitan: Monolaurate
   Form @ 25C: Liquid
   Saponification Value: 158-170
   Hydroxyl Value: 330-358
   Maximum Acid Value: 7
   HLB Value: 8.0

40:
   Sorbitan: Monopalmitate
   Form @ 25C: Flake
   Saponification Value: 140-150
   Hydroxyl Value: 275-305
   Maximum Acid Value: 7.5
   HLB Value: 6.5

60:
   Sorbitan: Monostearate
   Form @ 25C: Flake
   Saponification Value: 147-157
   Hydroxyl Value: 235-260
   Maximum Acid Value: 10
   HLB Value: 4.7

60K:
   Sorbitan: Monostearate
   Form @ 25C: Flake
   Saponification Value: 147-157
   Hydroxyl Value: 235-260
   Maximum Acid Value: 10
   HLB Value: 4.7

60KHM*:
   Sorbitan: Monostearate
   Form @ 25C: Flake
   Saponification Value: 147-157
   Hydroxyl Value: 235-260
   Maximum Acid Value: 10
   HLB Value: 4.7

* KHM denotes Kosher high melt point.

MAZER CHEMICALS: S-MAZ Sorbitan Fatty Acid Esters(Continued):

S-MAZ:

65K:
   Sorbitan: Tristearate
   Form @ 25C: Flake
   Saponification Value: 176-188
   Hydroxyl Value: 66-80
   Maximum Acid Value: 15
   HLB Value: 2.2

80:
   Sorbitan: Monooleate
   Form @ 25C: Liquid
   Saponification Value: 149-160
   Hydroxyl Value: 193-209
   Maximum Acid Value: 8
   HLB Value: 4.6

80K:
   Sorbitan: Monooleate
   Form @ 25C: Liquid
   Saponification Value: 149-160
   Hydroxyl Value: 193-209
   Maximum Acid Value: 8
   HLB Value: 4.6

83R:
   Sorbitan: Sesquioleate
   Form @ 25C: Liquid
   Saponification Value: 145-160
   Hydroxyl Value: 185-215
   Maximum Acid Value: 12
   HLB Value: 4.6

85:
   Sorbitan: Trioleate
   Form @ 25C: Liquid
   Saponification Value: 172-186
   Hydroxyl Value: 56-68
   Maximum Acid Value: 14
   HLB Value: 2.1

MAZER CHEMICALS: S-MAZ Sorbitan Fatty Acid Esters(Continued):

S-MAZ:

85K:
    Sorbitan: Trioleate
    Form @ 25C: Liquid
    Saponification Value: 172-186
    Hydroxyl Value: 56-68
    Maximum Acid Value: 14
    HLB Value: 2.1

90:
    Sorbitan: Monotallate
    Form @ 25C: Liquid
    Saponification Value: 145-160
    Hydroxyl Value: 180-210
    Maximum Acid Value: 10
    HLB Value: 4.3

95:
    Sorbitan: Tritallate
    Form @ 25C: Liquid
    Saponification Value: 168-186
    Hydroxyl Value: 55-85
    Maximum Acid Value: 15
    HLB Value: 1.9

MAZER CHEMICALS: T-MAZ Ethoxylated Sorbitan Fatty Acid Esters:

The T-MAZ Ethoxylated Sorbitan Fatty Acid Esters are emulsifiers with a wide range of hydrophilic characteristics. They are used individually or with the S-MAZ series of products to cover a wide range of oil in water and water in oil emulsification systems. Individually, they are excellent solubilizers of essential oils, wetting agents, viscosity modifiers, antistats, stabilizers and dispersing agents. They are used to prepare numerous products in the food, cosmetic, drug, textile and metalworking industries. Some T-MAZ Ethoxylated Sorbitan Fatty Acid Esters are also available for food use with Kosher Ⓤ Certification. They are designated with a "K" in the product name.

T-MAZ:

20:
    POE Content: 20
    Sorbitan: Monolaurate
    Form @ 25C: Liquid
    Saponification value: 40-50
    Hydroxyl Value: 96-108
    HLB Value: 16.7

28:
    POE Content: 80
    Sorbitan: Monolaurate
    Form @ 25: Liquid
    Saponification Value: 7-15
    Hydroxyl Value: 25-40
    HLB Value: 19.2

40:
    POE Content: 20
    Sorbitan: Monopalmitate
    Form @ 25C: Gel
    Saponification Value: 41-52
    Hydroxyl Value: 85-105
    HLB Value: 15.8

60:
    POE Content: 20
    Sorbitan: Monostearate
    Form @ 25C: Gel
    Saponification Value: 45-55
    Hydroxyl Value: 81-96
    HLB Value: 14.9

MAZER CHEMICALS: T-MAZ Ethoxylated Sorbitan Fatty Acid Esters (Continued):

T-MAZ:

60K:
POE Content: 20
Sorbitan: Monostearate
Form @ 25C: Gel
Saponification Value: 45-55
Hydroxyl Value: 81-96
HLB Value: 14.9

60KHM*:
POE Content: 20
Sorbitan: Monostearate
Form @ 25C: Gel
Saponification Value: 45-55
Hydroxyl Value: 81-96
HLB Value: 14.9

* KHM denotes Kosher high melt point

61:
POE Content: 4
Sorbitan: Monostearate
Form @ 25C: Paste
Saponification Value: 98-113
Hydroxyl Value: 170-200
HLB Value: 9.5

65K:
POE Content: 20
Sorbitan: Tristearate
Form @ 25C: Paste
Saponification Value: 88-98
Hydroxyl Value: 44-60
HLB Value: 10.5

80:
POE Content: 20
Sorbitan: Monooleate
Form @ 25C: Liquid
Saponification Value: 45-55
Hydroxyl Value: 65-80
HLB Value: 15.0

MAZER CHEMICALS: T-MAZ Ethoxylated Sorbitan Fatty Acid Esters (Continued):

T-MAZ:

80K:
   POE Content: 20
   Sorbitan: Monooleate
   Form @ 25C: Liquid
   Saponification Value: 45-55
   Hydroxyl Value: 65-80
   HLB Value: 15.0

80KLM**:
   POE Content: 20
   Sorbitan: Monooleate
   Form @ 25C: Liquid
   Saponification Value: 45-55
   Hydroxyl Value: 65-80
   HLB Value: 15.0
   ** KLM denotes Kosher low melt product

81:
   POE Content: 5
   Sorbitan: Monooleate
   Form @ 25C: Liquid
   Saponification Value: 96-104
   Hydroxyl Value: 134-150
   HLB Value: 10.0

85:
   POE Content: 20
   Sorbitan: Trioleate
   Form @ 25C: Liquid
   Saponification Value: 83-93
   Hydroxyl Value: 39-52
   HLB Value: 11.1

90:
   POE Content: 20
   Sorbitan: Monotallate
   Form @ 25C: Liquid
   Saponification Value: 45-55
   Hydroxyl Value: 65-80
   HLB Value: 14.9

95:
   POE Content: 20
   Sorbitan: Tritallate
   Form @ 25C: Liquid
   Saponification Value: 83-93
   Hydroxyl Value: 39-52
   HLB Value: 11.0

MILLIKEN CHEMICALS: SYN FACS

SYN FAC 222:
This aryl polyoxyether combines low foaming and high surface tension properties making it especially useful as a dispersant for latex paint and textile printing systems. It also offers advantages as a fiber finish component, a dispersant for degreasing solvents used in cleaning textile materials and metals; and as a scouring aid to improve crockfastness of naphthol dyed cellulosics. SYN FAC 222 can be used to produce anionic emuslifiers (phosphate or sulfate type).
Advantages:
* Good color acceptance
* Good color development
* Low foaming
* Outstanding heat stability, antistatic and emulsifying properties when used as a fiber finish component
* High surface tension
* Versatile--useful in many applications

General Characteristics:
Acid No.: 1
HLB: 11.5-12.5
Hydroxyl No.: 70-80
Ionic Nature: non-ionic

SYN FAC 334:
The SYN FAC 334 series are unique dispersant/emulsifiers for dispersing all types of pigments, insecticides, solvents, and cleaning compounds. The various hydrophile/lipophile balances of aryl ether to ethylene oxide provide efficient emulsifiers for multiple uses.
Advantages:
* Good color acceptance
* Good color development
* Good emulsifiers for aromatic solvents
* Low volatility
* Good color stability
* Low foaming characteristics

General Characteristics:
334:
Acid No.: 1
HLB: 11
Hydroxyl No.: 70-80
Ionic Nature: non-ionic
334-6:
Acid No.: 1
HLB: 9-10
Hydroxyl No.: 95-105
Ionic Nature: non-ionic
334-13:
Acid No.: 1
HLB: 13
Hydroxyl No.: 52-62
Ionic Nature: non-ionic

MILLIKEN CHEMICALS: SYN FACS(Continued):

SYN FAC 8210:

SYN FAC 8210 is a polyoxyaryl ether formulated to provide an efficient economical nonionic dispeersant for all types of pigments. It is also recommended in other applications where an excellent low foaming emulsifier-dispersant is required.

Benefits:
* Improved color yield in dispersing pigments
* Excellent dispersant for pigments, dyestuffs, titanium dioxide, and carbon black providing better mixing efficiency and dispersions with stable viscosity characteristics
* Excellent emulsification of aromatic and chlorinated solvents
* Low foaming properties allow easy processing
* Stable in both acid and alkaline system
* Excellent dispersant for degreasing solvents used in cleaning metals and textile materials

General Characteristics:
Acid No.: <1
HLB: 11.0
Hydroxyl No.: 45-58

SYN FAC 8216:

SYN FAC 8216 is a unique nonionic pigment dispersant designed for water based paints, coatings, inks and textile printing systems. SYN FAC 8216 is similar in structure to SYN FAC 222 and SYN FAC 334. Its principal difference is that it has a higher HLB and thus gives greater color development.

Advantages:
SYN FAC 8216 will produce much improved color development with most organic pigments than conventional nonionics such as nonyl or octyl phenol ethoxylates. In addition it is somewhat less foaming than conventional nonionics.

General Characteristics:
HLB: 15
Hydroxyl Number: 44-50
Cloud Point (1% Solution): 88C
Ionic Nature: Nonionic
Viscosity @ 50C: 100-300 cps
pH (5% solution): 6-8
Flash Point (C.O.C.): 300C
Weight Per Gallon: 9.2 lbs.
% Activity: 100%

MILLIKEN CHEMICALS: SYN FACS (Continued):

SYN FAC 8337:

SYN FAC 8337 is the potassium salt of a phosphated alkoxylated aryl phenol. Its primary use is in dispersing organic pigments; however, it is also recommended in other applications where low viscosity dispersions are required.

Benefits:
  * Gives more stable, low viscosity dispersions than conventional nonionic dispersants
  * Provides anionic characteristics to dispersants while maintaining the emulsifying efficiency of nonionic products
  * As a pound for pound replacement for sodium lauryl sulfate in pigment dispersions, SYN FAC 8337 offers improved stability, greater color value and lower foaming during mixing and printing
  * Excellent dispersant for titanium dioxide

General Characteristics:
  Appearance: Light yellow viscous liquid
  Chemical Form: 50% solution in water/alcohol
  HLB: >20
  Ionic Nature: Anionic
  Viscosity @ 25C: 1000 cps

SYN FAC TEA-97:

SYN FAC TEA-97 is an ethoxylated amine used as a solvent and/or dispersant for acidic materials. It is especially recommended for use in the dye industry for solubilizing and dyes, pigments and optical brighteners.

Advantages:
  * Plant proven
  * Applicable with wide range of pigments
  * Economical to use

General Characteristics:
  Acid Number: 22
  Color (Gardner): <10
  Hydroxyl Number: 610
  pH (10% of solution): 8.4
  Specific Gravity at 25C: 1.10
  Viscosity at 25C: 400 cps
  Water Content: <1

Application Information:
  In addition to its use in the dye industry, SYN FAC TEA-97 is very effective for dispersing soils in metal cleaning and related application. 1% concentration is adequate for most uses.

MOBAY CHEMICAL CORP.: Emulsifier K30:

Anionic Emulsifier for the polymerization of emulsions

Chemical Description: Mainly secondary sodium alkane sulphonates based on n-paraffin
Average Chain Length: approx. C15
Average molecular weight: approx. 330

Form supplied:
Emulsifier K30	95% flakes
Emulsifier K30	76% paste
Emulsifier K30	68% pumpable paste
Emulsifier K30	40% solution

Composition:

Emulsifier K30 95%:
  Active substance: approx. 95%
  Neutral oil: max. 1.2%

Emulsifier K30 76%:
  Active substance: approx. 76%
  Neutral oil: max. 1.0%

Emulsifier K30 68%:
  Active substance: approx. 68%
  Neutral oil: max. 0.9%

Emulsifier K30 40%:
  Active substance: approx. 40%
  Neutral oil: max. 0.6%

Application Fields:
  Polymerization emulsifier:
  Emulsifier K30 is used for the emulsion polymerization of various monomers to produce dispersions

  Emulsifier:
  Emulsifier K30 can be used because of its excellent emulsifying properties for processing natural latex, synthetic latices and plastic dispersions.

  Antistatic:
  Emulsifier K30 used as a polymerization emulsifier gives very good antistatic effect.

NL CHEMICALS: KELECIN F and 1081 Surfactants:

Fluid and Water Dispersible Modified Lecithin Surfactants

Description:
KELECIN are mixtures of soybean organic phosphatides and triglycerides having unusual surface active properties. KELECIN F (fluid grade) is a relatively low viscosity liquid for convenience in handling. KELECIN 1081 has water dispersibility for use in water-thinned products.

Suggested Uses and Performance:
* Cosmetics--emulsifying agent and emollient for skin creams, lotions, shampoos, and hair tonic; improves detergency of soaps.
* Food Products--antispattering agent for frying; blending aid for dry and fluid emulsifier use ingredients in chocolates; dispersant/emulsifier for flavorings, yeast, flours; browning agent in baked goods; rancidity inhibitor in lards, margarines and peanut butter; aid to moisture retention and texture in whips, ice creams, baked goods.

Tentative Specifications:

KELECIN F:
   Viscosity, Stokes, approx.: 150
     Gardner-Holdt, max: Z6
   Color, Gardner, max.: 10
     Pounds/Gallon: 8.45-8.55
   Specific Gravity, 25C, approx.: 1.02
   Acid Value, max.: 32.0
   Acetone Insoluble, % min.: 62.0
   Benzene Insoluble, % max.: 0.3
   Moisture, % Max.: 1.0

KELECIN 1081:
   Viscosity, Stokes, approx.: 25-60
     Gardner-Holdt, max.: Z2
   Color, Gardner, max.: 12
     Pounds/Gallon: 8.53-8.63
   Specific Gravity, 25C, approx: 1.03
   Acid Value, max.: 22-31
   Acetone Insoluble, % min.: 49.0

PATCO PRODUCTS: ALPHADIM 90AB Distilled Monoglyceride:

Description:
ALPHADIM 90AB is a high purity, molecularly distilled monoglyceride prepared from fully hardened edible fats and glycerine.

FDA Clearance & Labeling:
ALPHADIM 90AB is a G.R.A.S. food ingredient and may be used at the level required to perform the intended function. It may be labeled in a finished food as "monoglyceride."

Applications & Use Levels:
ALPHADIM 90AB is an emulsifying and starch complexing agent which provides significant functional improvements in processing and storage stability for a wide variety of food products.

Applications: Table Margarine
Typical Use Levels: 0.1-0.3% formula wt.
Function: Provides a stable, finely dispersed emulsion, prevents moisture loss ("weeping").

Applications: Coffee Whitener
Typical Use Levels: 0.1-1.0% formula wt.
Function: Stabilizes and disperses fat particles, improves whitening.

Applications: Pasta, Snacks, Cereal
Typical Use Levels: 0.5-1.75% dry wt.
Function: Prevents stickiness, improves texture, reduces scorching.

Applications: Dehydrated Potatoes
Typical Use Levels: 0.3-0.8% dry wt.
Function: Improves texture in reconstituted potato products, prevents stickiness during dehydration.

Applications: Bread, Rolls
Typical Use Levels: 0.2-0.5% flour wt.
Function: Extends shelf life, improves softness.

Applications: Confection Coatings, Chewing Gum
Typical Use Levels: 0.5-1.0% formula wt.
Function: Prevents stickiness, improves cutting and mouth release

Applications: Peanut Butter
Typical Use Levels: 1.0-2.5% formula wt.
Function: Eliminates oil separation, reduces stickiness

Typical Properties:
ALPHADIM 90AB is a white to cream colored fine bead.
Alpha Monoester Content: 90% Min.
Iodine Value: 3 Max.
Saponification Value: 150-165

PATCO PRODUCTS: ALPHADIM 90NLK Distilled Monoglyceride:

Description:
ALPHADIM 90NLK is a kosher certified, high purity, molecularly distilled monoglyceride prepared from refined sunflower oil and glycerine. TBHQ and citric acid have been added to ensure product quality.

FDA Clearance & Labeling:
ALPHADIM 90NLK is a G.R.A.S. food ingredient and may be used at the level required to perform the intended function. It may be labeled in a finished food as "monoglyceride."

Applications & Use Levels:
ALPHADIM 90NLK provides a stable, finely dispersed emulsion in diet margarines or low calorie spreads. For shortening, improved creaming properties result from use of ALPHADIM 90NLK.

Applications: Margarines/Spreads
Typical Use Levels: 0.3-6.0% formula wt.
Function: Stable, finely dispersed emulsion, prevents moisture loss ("weeping")

Applications: Cake Shortenings
Typical Use Levels: 0.3-0.6% formula wt.
Function: Uniform fat dispersion in batter, improved volume and tenderness

Applications: Icing Shortenings
Typical Use Levels: 0.3-1.0% formula wt.
Function: High water absorption, creamy texture

Typical Properties:
ALPHADIM 90NLK is a white to pale yellow colored fluid plastic material at room temperature.
  Alpha Monoester Content: 90% Min.
  Iodine Value: 110 Approx.
  Free Glycerine: 1.5% Max.
  Saponification Value: 150-165
  Melting Point: 40C Approx.

PATCO PRODUCTS: ALPHADIM 90SBK:

Description:
ALPHADIM 90SBK is a kosher certified, high purity, molecularly distilled monoglyceride prepared from fully hardened soybean oil and glycerine.

FDA Clearance:
ALPHADIM 90SBK is a G.R.A.S. food ingredient and may be used at the level required to perform the intended function. It may be labeled in a finished food as "monoglyceride."

Applications: Table Margarine
Typical Use Levels: 0.1-0.3% formula wt.
Function: Stable, finely dispersed emulsion, prevents moisture loss ("weeping").

Applications: Coffee Whitener
Typical Use Levels: 0.1-1.0% formula wt.
Function: Stabilize and disperse fat particles, improve whitening

Applications: Pasta, Snacks, Cereal
Typical Use Levels: 0.5-1.75% dry wt.
Function: Prevents stickiness, improve texture, reduce scorching

Applications: Dehydrated Potatoes
Typical Use Levels: 0.3-0.8% dry wt.
Function: Starch complexing to improve texture in reconstituted potato products. Prevents stickiness during dehydration

Applications: Bread, Rolls
Typical Use Levels: 0.2-0.5% flour wt.
Function: Starch complexing, softening, extend shelf life.

Applications: Confection Coatings, Chewing Gum
Typical Use Levels: 0.5-1.0% formula wt.
Function: Prevent stickiness, improve cutting and mouth release, improve gloss with fine crystal structure.

Applications: Peanut Butter
Typical Use Levels: 1.0-2.5% formula wt.
Function: Eliminate oil separation, reduce stickiness, improve flavor and high temperature stability for processing and storage.

Typical Properties:
ALPHADIM 90SBK is a white to cream colored fine bead.
Alpha Monoester Content: 90% Min.
Iodine Value: 3 Max.
Free Glycerine: 1.5% Max.
Saponification Value: 150-165
Melting Point: 72C Approx.

PATCO PRODUCTS: DOUBLE SOFT Concentrated Hydrate:

Description:
DOUBLE SOFT is a highly stable, hydrated emulsion of high purity, molecularly distilled monoglyceride. Propionic and/or acetic acid has been added to ensure product quality.

FDA Clearance & Labeling:
DOUBLE SOFT is a G.R.A.S. food ingredient and may be used at the level required to perform the intended function. It may be labeled in a finished food as "monoglyceride."

Applications & Use Levels:
DOUBLE SOFT is recommended for use as a crumb softener in yeast raised products. It is concentrated and should be used at one-half the level of ordinary hydrates.

Typical Properties:
DOUBLE SOFT is an ivory colored semi-solid.
Alpha Monoester Content: 40% Min.
Volatiles: 57% Max.
Saponification Value: 66-78
pH (2%): 3.6-4.2

VERV
In Bakery Products

Description:
VERV calcium stearoyl-2-lactylate is 100% active and is the reaction product of stearic acid and lactic acid, neutralized to the calcium salt.

Analytical Characteristics:
Form: free flowing powder
Odor: mild, caramel-like
Color: light tan
Acid Number: 50 to 86
Calcium Content: 4.2 to 5.2%
Ester Number: 125 to 164

Applications and Use Levels:
VERV is designed for use as a dough conditioner and softener in yeast leavened bakery products and mixes at a level of 0.5%.
VERV can be used in whipped toppings at a level up to 0.3% of the weight of the finished topping.
VERV is also used as a conditioning agent in dehydrated potatoes at a level up to 0.5% of the weight of the dehydrated potatoes.

PATCO PRODUCTS: EMPLEX Sodium Stearoyl Lactylate:

Material Description:
EMPLEX sodium stearoyl lactylate is manufactured by the reaction of stearic acid and lactic acid, neutralized to the sodium salt. A kosher-certified grade is available.

FDA Clearance & Labeling:
EMPLEX is approved for use in foods as described in 21 CFR 172.846. It may be labeled as "sodium stearoyl lactylate."

Applications & Use Levels:
EMPLEX is a starch and protein complexing agent used in bakery products. It is also used as an emulsifier and conditioning agent in a variety of processed foods. EMPLEX is readily dispersible in hot oil or water.

Applications: Bread
Typical Use Levels: 4-8 oz./cwt flour
Function: Improves loaf volume, texture, and softness.

Applications: Buns, Rolls
Typical Use Levels: 6-8 oz./cwt flour
Function: Improves dough flow, slicing, volume, texture, softness

Applications: Cheese Substitutes & Imitation Cheese
Typical Use Levels: 0.1-0.2% formula wt.
Function: Emulsifies and stabilizes fat, provides excellent body

Applications: Dehydrated Potatoes
Typical Use Levels: 0.3-0.5% dry wt.
Function: Improves sheeting and release, provides tolerance in rehydration, reduces stickiness.

Applications: Non-Dairy Coffee Whiteners
Typical Use Levels: 0.1-0.3% formula wt.
Function: Emulsifies and stabilizes fat; improves whitening.

Applications: Puddings & Snack Dips
Typical Use Levels: 0.1-0.2% formula wt.
Function: Stabilizes emulsion, improves texture and appearance.

Applications: Sauces & Gravies
Typical Use Levels: 0.1-0.25% formula wt.
Function: Emulsifies and stabilizes fat, reduces "skinning"

Applications: Whipped Toppings & Fillings
Typical Use Levels: 0.1-0.2% formula wt.
Function: Emulsifies fat; aids aeration; improves texture.

Typical Properties:
EMPLEX is a light tan powder with a mild caramel odor.
Acid Value: 60-80
Ester Number: 150-190
Sodium Content: 3.5-5.0%

PATCO PRODUCTS: STARPLEX 90 Water Dispersible Distilled Monoglyceride:

Description:
STARPLEX 90 is a high purity, molecularly distilled monoglyceride prepared from edible fats or oils and glycerine. BHA and citric acid have been added to retain freshness. A kosher-certified version is available.

FDA Clearance & Labeling:
STARPLEX 90 is a G.R.A.S. food ingredient and may be used at the level required to perform the intended function. It may be labeled in a finished food as "monoglyceride."

Applications & Use Levels:
STARPLEX 90 is a concentrated monoglyceride which hydrates on contact with water. It provides increased hydration and improved functionality of the monoglyceride in new and existing food systems. Best results are achieved when the melting point of STARPLEX 90 is not exceeded.

Applications: White Bread or Buns
Typical Use Levels: 2-6 oz./cwt flour
Function: Improves softness, extends shelf life.

Applications: Variety Breads
Typical Use Levels: 2-4 oz./cwt flour
Function: Improves softness, extends shelf life.

Applications: English Muffins
Typical Use Levels: 2-6 oz./cwt flour
Function: Extends shelf life, prevents excessive toughness and chewiness.

Applications: Sweet Goods
Typical Use Levels: 4-6 oz./cwt flour
Function: Improves softness, extends shelf life.

Applications: Sauces and Gravies
Typical Use Levels: 0.2-0.5% formula wt.
Function: Emulsifies and stabilizes fat, reduces "skinning," improves cling and freeze/thaw stability.

Applications: Pasta
Typical Use Levels: 0.5-1.0% dry wt.
Function: Improves tolerance and freeze/thaw stability, increases firmness, minimizes stickiness and clumping.

Applications: Cereal
Typical Use Levels: 0.5-1.0% dry wt.
Function: Reduces stickiness, improves extrusion and texture.

Typical Properties:
  STARPLEX 90 is a white powder.
  Alpha Monoester Content: 90% Min.
  Free Glycerine: 1.5% Max.
  Saponification Value: 150-165

QUANTUM CHEMICAL CORP.: EMEREST Glycerol Esters:

EMEREST Glycerol Esters are based on fatty acids. In the leather and textile industries, they are used as components in lubricants, softeners, and dye carriers, and as co-emulsifiers for synthetic fiber spin finishes. Industrial applications include uses as lubricants, rust preventatives, and mold release agents.

EMEREST 2400 Glycerol Monostearate is used as a component in hand creams and other cosmetic formulations. This versatile ester is also a component for industrial lubricants and lubricant softeners for textiles.

HLB: 3.9
Form @ 25C: Solid M.P. 58
Viscosity cSt 100F: Solid

EMEREST 2401 Glycerol Monostearate is the technical grade of EMEREST 2400 having similar properties and functions. This product is not recommended for applications where color is a critical parameter.

HLB: 3.9
Form @ 25C: Solid M.P. 58
Viscosity cSt 100F: Solid

EMEREST 2407 Glycerol Monostearate, SE differs from the two previous products with respect to acid value. EMEREST 2407 has an acid value of approximately 20, and is self-emulsifying. It is used in various cosmetic creams and lotions.

HLB: 5.1
Form @ 25C: Solid M.P. 58
Viscosity cSt 100F: Solid

EMEREST 2410 Glycerol Monoisostearate is a liquid ester derived from isostearic acid that exhibits excellent oxidation and color stability. It has unique emollient, lubricating and w/o emulsification properties.

HLB: 2.9
Form @ 25C: Liquid Pour Pt. 5
Viscosity cSt 100F: 260

QUANTUM CHEMICAL CORP.: EMEREST Glycerol Esters(Continued):

EMEREST 2421 Glycerol Monooleate has a monoester content of 50-60%, free glycerol 1%, and moisture 1%, maximum. It is used as a component in mold release agents, a vehicle for agricultural insecticides, an anti-icing fuel additive, and a rust preventive additive for compounded oils. It is used in the textile industry as a lubricant component in synthetic fiber spin finishes.

HLB: 3.4
Form @ 25C: Liquid Pour Pt. 19
Viscosity cSt 100F: 91

EMEREST 2419 Glycerol Dioleate can be used as a very low HLB emulsifier with high oil solubility. Applications in water-in-oil emulsions or industrial lubricants are suggested.

HLB: 1.6
Form @ 25C: Liquid Pour Pt. <0
Viscosity cSt 100F: 55

EMEREST 2423 Glycerol Trioleate (GTO) is sometimes called synthetic olive oil. When emulsified, it is an excellent lubricant for metals, leather and textiles. Sulfated GTO is used as a softener in the leather and textile industries.

HLB: 0.6
Form @ 25C: Liquid Pour Pt. 9
Viscosity cSt: 43

EMEREST 2452 Triglycerol Diisostearate is an excellent solvent for dyes and pigments with improved oxidative stability. It can be directly substituted for castor oil to wet pigment more rapidly for faster dispersions.

HLB: 6.7
Form @ 25C: Liquid Pour Pt. 4
Viscosity cSt: 990

QUANTUM CHEMICAL CORP.: EMID Alkanolamides:

EMID Alkanolamides are based on coconut and other selected fatty acids. They are used as foam boosting and stabilizing agents and as thickeners in a variety of formulations including shampoos, detergents, bubble baths, rug cleaners and general purpose industrial cleaners.

EMID 6500 Coconut Monoethanolamide is a foam stabilizer and thickener for shampoos, liquid detergents and rug cleaners.
   Form @ 25C: Flaked Solid M.P. 72

EMID 6590 Lauramide DEA (and) Propylene Glycol is a viscosity builder and foam stabilizer containing propylene glycol for ease of handling in heavy duty liquid detergent formulations, shampoos and bath products.
   Form @ 25C: Liquid Pour Pt. <-10
   Viscosity cSt 100F: 176

EMID 6514 Coconut Super Diethanolamide is an easy-to-handle foam stabilizer, thickener and detergent component for various liquid cleaning compounds.
   Form @ 25C: Liquid Pour Pt. 10
   Viscosity cSt 100F: 336

EMID 6515 Coconut Super Diethanolamide boosts and stabilizes foam, and inhibits redeposition of soils.
   Form @ 25C: Liquid Pour Pt. 0
   Viscosity cSt 100F: 390

EMID 6545 Oleic Diethanolamide is used as an emulsifier for mineral oils in formulating antistatic fiber processing aids and yarn lubricants. EMID 6545 also serves as a foam suppressant in dye carrier and solvent emulsions.
   Form @ 25C: Liquid Pour Pt. -3
   Viscosity cSt 100F: 290

EMID 6533 modified Coconut Diethanolamide is an emulsifier, thickening agent and moderate foamer for controlled suds detergents. In the textile industry it is used as a fulling and scouring agent.
   Form @ 25C: Liquid Pour Pt. -15
   Viscosity cSt 100F: 345

EMID 6521 Modified Coconut Diethanolamide is a thickener, foam booster, stabilizer and emulsifier.
   Form @ 25C: Liquid Pour Pt. <25
   Viscosity cSt 100F: 614

EMID 6529 Modified Coconut Diethanolamide is a thickener, foam booster, stabilizer and emulsifier.
   Form @ 25C: Liquid Pour Pt. <25
   Viscosity cSt 100F: 619

QUANTUM CHEMICAL CORP.: EMSORB Ethoxylated Sorbitan Esters:

The EMSORB Ethoxylated Sorbitan Esters are hydrophilic surfactants that are frequently used as emulsifiers in blends with the unethoxylated lipophilic esters from which they were derived, i.e., with the EMSORB Sorbitan Fatty Acid Esters. They are frequently used as the principal emulsifier for emulsifiable industrial processing and finishing oils. Additionally, these type products are established as important textile chemicals where they are used as fiber antistats, fiber-yarn lube components, and as functional emulsifiers in fabric finishing softeners.

Product Name:

EMSORB 6900 POE (20) Sorbitan Monooleate, a hydrophilic emulsifier, functions as a co-emulsifier for petroleum oils, fats, solvents and waxes in household products, industrial lubricants and textile dye carriers. It is also used as a dispersant for pigments in coatings, a solubilizer for oils and fragrances, and as an emulsifier for aliphatic alcohols in tobacco sucker control concentrates.

HLB: 15.0
Form @ 25C: Liquid Pour Pt. -12
Viscosity cSt 100F: 200

EMSORB 6901 POE (5) Sorbitan Monooleate, an effective O/W emulsifier and lubricant, is used with mineral and vegetable oils in the formulation of industrial lubricants. Textile oils emulsified with EMSORB 6901 yield excellent lubricants and softeners for fibers and yarns.

HLB: 10.0
Form @ 25C: Liquid Pour Pt. -15
Viscosity cSt 100F: 210

EMSORB 6903 POE (20) Sorbitan Trioleate is an oil in water emulsifier for petroleum oils, fats, waxes and alkyl esters. It is an excellent lubricant for metals, textiles, and leather, and is used as an emulsifier/lubricant in soluble oils for metal processing and finishing. EMSORB 6903 also serves as a lubricant and highly efficient emulsifier for oils used in textile processing and finishing, glass fiber lubricants, and automotive lubricant additives.

HLB: 11.1
Form @ 25C: Liquid Pour Pt. -15
Viscosity cSt 100F: 160

QUANTUM CHEMICAL CORP.: EMSORB Ethoxylated Sorbitan Esters
(Continued):

Product Name:

EMSORB 6917 POE (16) Sorbitan Trioleate is similar to EMSORB 6903 but is slightly less hydrophilic.
HLB: 10.0
Form @ 25C: Liquid Pour Pt. <-10
Viscosity cSt 100F: 122

EMSORB 6905 POE (20) Sorbitan Monostearate, a waxy semi-solid, functions as an O/W emulsifier for mineral oil, fats, and waxes. It is a good fiber-to-metal textile lubricant and is used as a co-emulsifier with EMSORB 2505 in paraffin wax emulsions for textiles, paper and wallboard coatings. It is also used as an emulsifier in household products.
HLB: 15.2
Form @ 25C: Liquid Pour Pt. 25
Viscosity cSt 100F: 250

EMSORB 6906 POE (3) Sorbitan Monostearate is a W/O emulsifier which is used in household formulations. It also functions as a fiber-to-metal lubricant for synthetic and cellulosic fibers and yarns.
HLB: 9.0
Form @ 25C: Solid M.P. 42

EMSORB 6908 POE (16) Sorbitan Tristearate is a waxy O/W emulsifier for petroleum oils and natural fats. Because of its lubricating and softening properties, it is useful as a component in textile processing and finishing compounds. Its balanced oil and water solubility permit it to be used as a primary emulsifier.
HLB: 10.0
Form @ 25C: Solid M.P. 38

EMSORB 6909 POE (4) Sorbitan Monostearate is a hydrophobic emulsifier which is a waxy solid at room temperature.
HLB: 4.0
Form @ 25C: Solid M.P. 35

EMSORB 6915 POE (20) Sorbitan Monolaurate is a versatile O/W emulsifier and solubilizer of petroleum oils, solvents, and fats. High water solubility enhances its ability to solubilize petroleum solvents. In the textile industry it is used as an emulsifier for dye carriers, as an antistatic scrooping agent in primary spin finishes, and as a fiber processing aid.
HLB: 16.7
Form @ 25C: Liquid Pour Pt. -10
Viscosity cSt 100F: 160

QUANTUM CHEMICAL CORP.: EMSORB Sorbitan Fatty Acid Esters:

EMSORB Sorbitan Fatty Acid Esters are lipophilic emulsifiers and coupling agents serving as integral components of emulsifiable industrial mold release agents, textile fiber and yarn lubricants, and textile finishing softeners. They are widely used as surfactants and functional components in many kinds of industrial and household products.

Product Name:

EMSORB 2500 Sorbitan Monooleate is an oil soluble emulsifier, coupling agent, lubricant, and softener for textile fibers and leather. It can be used with water soluble emulsifiers in formulating clear emulsifiable concentrates of petroleum oils and waxes, natural fats and waxes, and alkyl esters. Small amounts will usually clarify (couple) such concentrates. EMSORB 2500 is also widely used in cosmetics and household products.

HLB: 4.6
Form @ 25C: Liquid Pour Pt. <0
Viscosity cSt 100F: 360

EMSORB 2502 Sorbitan Sesquioleate is a primary emulsifier for water in oil systems and an effective co-emulsifier for oil in water systens for mineral oil, fats, and waxes. It is also a versatile coupling agent for water soluble materials in O/W systems. Suggested applications include its use as a W/O emulsifier for consumer household aerosols and as an O/W co-emulsifier for cosmetics, industrial oils and textile oils.

HLB: 4.5
Form @ 25C: Liquid Pour Pt. <0
Viscosity cSt 100F: 475

EMSORB 2503 Sorbitan Trioleate is similar to EMSORB 2500, but is more hydrophobic. Its oil solubility makes it a very effective coupling agent and co-emulsifier for mineral oil. EMSORB 2503 is used in the processing and softening of textiles and leather.

HLB: 2.1
Form @ 25C: Liquid Pour Pt. <0
Viscosity cSt 100F: 100

QUANTUM CHEMICAL CORP.: EMSORB Sorbitan Fatty Acid Esters (Continued):

Product Name:

EMSORB 2505 Sorbitan Monostearate, a hydrophobic emulsifier, is used to produce emulsions of mineral oils, fats, waxes and silicones. Suggested applications as a co-emulsifier include industrial oils, household products and cosmetics. Used in conjunction with EMSORB 6905 POE (20) Sorbitan Monostearate, this system finds application in paraffin wax emulsions and silicone defoamers for processing paper and textiles. As a textile lubricant, it helps reduce fiber-to-metal friction.

HLB: 5.2
Form @ 25C: Solid M.P.: 50

EMSORB 2510 Sorbitan Monopalmitate, an oil soluble waxy emulsifier, is used in cosmetic and household products and is a superior fiber-to-metal lubricant for synthetic and cellulosic fibers.

HLB: 6.5
Form @ 25C: Solid M.P. 47

EMSORB 2515 Sorbitan Monolaurate is a water dispersible emulsifier which exhibits anti-foam properties. Suggested uses include emulsifier systems for oils, fats, and solvents in household specialties, industrial oils, cosmetics and emulsion polymerization

HLB: 8.0
Form @ 25C: Liquid Pour Pt. 15
Viscosity cSt: 1000

EMSORB 2516 Sorbitan Monoisostearate is an effective auxiliary emulsifier. It is almost odor-free and is very light in color.

HLB: 4.6
Form @ 25C: Solid M.P. 50
Viscosity cSt: 1200

EMSORB 2518 Sorbitan Diisostearate is an effective auxiliary emulsifier. It is nearly odor-free and is very light in color.

HLB: 3.0
Form @ 25C: Liquid Pour Pt. -4
Viscosity cSt: 730

QUANTUM CHEMICAL CORP.: Ethoxylated Fatty Acids and Polyethylene Glycol Fatty Acid Esters:

Quantum's ethoxylated fatty acids and polyethylene glycol (PEG) fatty acid esters are nonionic, specialized, mono and diesters of various fatty acids. These products, made either by ethoxylation or by esterification with polyethylene glycol, have a wide range of surfactant properties. They provide lubricity and softening as components in a variety of formulated industrial and textile processing oils, and are useful as functional emulsifiers in textile fabric finishing softeners. They are generally strong emulsifiers of fats, oils and solvents, and are widely used as emulsifiers in textile, industrial and household products, as well as in cosmetics, pharmaceuticals and other high-purity end uses. A number of these products meet the requirements for use as indirect food additives under conditions described in CFR Title 21.

Product Name:

EMEREST 2634 PEG 300 Monopelargonate is a water dispersible surfactant whose primary end-use is an overspray for open-end spinning of synthetic fibers.

HLB: 2.8
Form @ 25C: Liquid Pour Pt. <-15
Viscosity cSt 100F: 25

EMEREST 2654 PEG 400 Monopelargonate is a water-soluble surfactant used as a base lubricant for synthetic fiber spin finishes, open-end spinning, and overspray finishes. It is also an excellent co-emulsifier and coupling agent for many formulations.

HLB: 14.3
Form @ 25C: Liquid Pour Pt. 5
Viscosity cSt 100F: 34

EMEREST 2620 PEG 200 Monolaurate can be used as a nonionic emulsifier or coupling agent. It is used as a defoamer in water base coatings, a viscosity depressant in vinyl plastisols, a viscosity control additive in hair rinse formulations, and as a paper softener.

HLB: 9.3
Form @ 25C: Liquid Pour Pt. 9
Viscosity cSt 100F: 123

QUANTUM CHEMICAL CORP.: Ethoxylated Fatty Acids and Polyethylene Glycol Fatty Acid Esters(Continued):

Product Name:

EMEREST 2630 PEG 300 Monolaurate is a moderately hydrophilic emulsifier which is primarily used as a lubricant component and scrooping agent for textile fibers and yarn, applied either from aqueous or emulsifiable mineral oil systems. It also functions as a viscosity control agent for plastisols.

HLB: 12.1
Form @ 25C: Liquid Pour Pt. 9
Viscosity cSt 100F: 37

EMEREST 2650 PEG 400 Monolaurate functions as a wetting agent, defoamer, and leveling agent in latex paints. This hydrophilic surfactant is also used as a dispersant in pigment grinding, as a solubilizer for oils and solvents, and as an anti-block agent in vinyls. In textile fiber processing, it imparts fiber-to-fiber cohesion and fiber-to-metal lubrication.

HLB: 13.2
Form @ 25C: Liquid Pour Pt. 12
Viscosity cSt 100F: 41

EMEREST 2661 PEG Monolaurate functions as a water-soluble lubricant in processing synthetic fibers.

HLB: 14.8
Form @ 25C: Liquid Pour Pt. 14
Viscosity cSt 100F: 60

EMEREST 2622 PEG 200 Dilaurate, an oil-soluble surfactant, can be used as a co-emulsifier and lubricant in self-emulsifiable textile and industrial oils, as a mold release agent, and as a viscosity control agent. EMEREST 2622 is also used in specialty paper coatings.

HLB: 7.6
Form @ 25C: Liquid Pour Pt. 0
Viscosity cSt 100F: 22

EMEREST 2652 PEG 400 Dilaurate, an oil-soluble emulsifier, can be used as a softener, lubricant, and release agent for paper. In the textile industry it serves as a coupler and lubricant in synthetic fiber spin finishes.

HLB: 10.8
Form @ 25C: Liquid Pour Pt. 8
Viscosity cSt 100F: 38

QUANTUM CHEMICAL CORP.: Ethoxylated Fatty Acids and Polyethylene Glycol Fatty Acid Esters(Continued):

Product Name:

TRYDET 2685 Ethoxylated Fatty Acid is an emulsifier for mineral oils and fats. It serves as an excellent lubricant.
HLB: 7.5
Form @ 25C: Solid M.P. 37

TRYDET 2670 POE (5) Stearic Acid is an emulsifier for mineral oils and fats. It is used as a lubricant in the leather industry and as a napping softener in the textile industry, providing scroop and fiber-to-metal lubricity. TRYDET 2670 is also used in mineral oil emulsions for polishes and lard oil emulsions for metal buffing compounds, and as as an O/W emulsifier for air fresheners.
HLB: 9.2
Form @ 25C: Liquid Pour Pt. 25
Viscosity cSt 100F: 44

TRYDET 2636 POE (7) Stearic Acid is a waxy emulsifier for oils and fats in industrial lubricants. It imparts softening and lubricating properties to textiles and leather. TRYDET 2636 is very similar to TRYDET 2670, but has better water dispersibility.
HLB: 10.1
Form @ 25C: Solid M.P. 28
Viscosity cSt 100F: 57

EMEREST 2640 PEG 400 Monostearate functions as an emulsifier of oils and fats in the manufacture of fluid or paste emulsions for industrial lubricants, consumer products, and texile lubricants and softeners. This waxy lipophilic surfactant is also used as a thickening agent and stabilizer for starch coatings on paper, and as a water dispersible paper size. EMEREST 2640 is also a good lubricant for channeling wire through conduit.
HLB: 12.0
Form @ 25C: Solid M.P. 32
Viscosity cSt 100F: 57

EMEREST 2662 PEG 600 Monostearate is an emulsifier for industrial and textile formulations and can be used as a viscosity modifier. Textile applications include its use as a component in softeners and self-emulsifying lubricants.
HLB: 13.8
Form @ 25C: Solid M.P. 40

QUANTUM CHEMICAL CORP.: Ethoxylated Fatty Acids and Polyethylene Glycol Fatty Acid Esters(Continued):

Product Name:

EMEREST 2610 PEG 1000 Monostearate is a water soluble emulsifier used to emulsify glycerol monostearate in nonionic textile lubricants and softeners. EMEREST 2610 is used as an emulsifier and thickener. In starch solutions, it acts as an antigellant.

HLB: 15.7
Form @ 25C: Solid M.P. 36

TRYDET 2672 POE (40) Stearic Acid is a strongly hydrophilic emulsifier, stabilizer, antigellant, and lubricant. It is used as an emulsifier for glycerol monostearate and other waxy esters in the production of concentrated, pourable textile lubricants and softeners, and as a stabilizer and antigellant for starch solutions. Other applications include its use as a nonabrasive coating for glass bottles.

HLB: 17.3
Form @ 25C: Solid M.P. 50

EMEREST 2675 POE (50) Stearic Acid (30% aqueous) is a very hydrophilic emulsifier used for preparing solubilized oils. It is also used as a viscosity modifier and as a softener or plasticizer in acrylic or vinyl resin emulsions.

HLB: 17.8
Form @ 25C: Liquid Pour Pt. 0
Viscosity cSt 100F: 671

EMEREST 2642 PEG 400 Distearate is a lipophilic waxy surfactant that performs as an emulsifier and thickener in cosmetic and industrial emulsions.

HLB: 7.5
Form @ 25C: Solid M.P. 36
Viscosity cSt 100F: 52

EMEREST 2625 PEG 200 Monoisostearate exhibits unique surfactant and lubricant properties resulting from the liquid, long-chain, saturated fatty acid from which it is made. It can be used as a component in fiber lubricants, textile fiber processing aids and concentrated liquid fabric softeners.

HLB: 8.3
Form @ 25C: Liquid Pour Pt -8
Viscosity cSt 100F: 50

QUANTUM CHEMICAL CORP.: Ethoxylated Fatty Acids and Polyethylene Glycol Fatty Acid Esters(Continued):

Product Name:

TRYDET 2644 PEG 400 Monoisostearate is similar in end uses to EMEREST 2625 but is more hydrophilic.

HLB: 11.3
Form @ 25C: Liquid Pour Pt. 10
Viscosity cSt 100F: 70

EMEREST 2624 PEG Monooleate is an oil-soluble emulsifier for mineral oils, fatty oils, and solvents. It is used as a cutting oil emulsifier, a solvent emulsifier in metal cleaners and degreasers, a W/O emulsifier for consumer pesticide aerosols, and as a lubricant component in textile processing. It also softens and lubricates leather during tanning.

HLB: 8.3
Form @ 25C: Liquid Pour Pt. <-15
Viscosity cSt 100F: 34

EMEREST 2632 PEG 300 Monooleate is an emulsifier and lubricant with properties and applications similar to EMEREST 2624. It is used as a self-emulsifying component in formulating textile softeners.

HLB: 10.4
Form @ 25C: Liquid Pour Pt -10
Viscosity cSt 100F: 46

EMEREST 2646 PEG 400 Monooleate, a moderately hydrophilic emulsifier and lubricant, is used as an emulsifier for solvents in pesticide carriers and metal cleaners. In the textile industry it is used in the formulation of specialty detergents and dyeing assistants. It is also used as an emulsifier for neatsfoot oil in leather fat liquoring and as a rewetting agent for paper.

HLB: 11.8
Form @ 25C: Liquid Pour Pt. 5
Viscosity cSt 100F: 52

TRYDET 2676 POE (10) Oleic Acid, a moderately hydrophilic emulsifier and lubricant, is used as an emulsifier for solvents in pesticide carriers and metal cleaners. In the textile industry it is used in the formulation of specialty detergents and dyeing assistants. It is also used as an emulsifier for neatsfoot oil in leather fat liquoring and as a rewetting agent for paper.
HLB: 12.2
Form @ 25C: Liquid Pour Pt. 14
Viscosity cSt 100F: 54

QUANTUM CHEMICAL CORP.: Ethoxylated Fatty Acids and Polyethylene Glycol Fatty Acid Esters(Continued):

Product Name:

EMEREST 2660 PEG 600 Monooleate is a water-soluble emulsifier and detergent. It functions as an emulsifier in specialty lubricants, as a dye leveling agent in textiles, and as a major component in systems used for the acid washing of printed circuit boards.

HLB: 13.6
Form @ 25C: Liquid Pour Pt. 18
Viscosity cSt 100F: 75

EMEREST 2618 PEG 4000 Monooleate is a water soluble emulsifier.

HLB: 18.8
Form @ 25C: Solid M.P. 55

EMEREST 2617 PEG 6000 Monoleate is a strongly hydrophilic emulsifier, stabilizer and lubricant.

HLB: 19.2
Form @ 25C: Solid M.P. 58

EMEREST 2619 PEG 6000 Monooleate (50%) is similar to EMEREST 2617 but is easier to handle.

HLB: 19.2
Form @ 25C: Liquid Pour Pt. 0
Viscosity cSt 100F: 1600

EMEREST 2647 PEG Sesquioleate finds application as an oil and fat emulsifier in industrial and textile lubricants.

HLB: 9.4
Form @ 25C: Liquid Pour Pt. -6
Viscosity cSt @ 100F: 50

EMEREST 2648 PEG 400 Dioleate is a lipophilic liquid emulsifier and solubilizer for mineral oils, fats, and solvents. A prime application for EMEREST 2648 is the emulsification of kerosene in agricultural and pesticide sprays. EMEREST 2648 is also used in the emulsification of latex paints, metalworking fluids, solvents, and specialty and industrial lubricants.

HLB: 8.8
Form @ 25C: Liquid Pour Pt. -6
Viscosity cSt 100F: 45

QUANTUM CHEMICAL CORP.: Ethoxylated Fatty Acids and Polyethylene Glycol Fatty Acid Esters(Continued):

EMEREST 2665 PEG 600 Dioleate is slightly more hydrophilic than EMEREST 2648, functioning in many similar applications.

HLB: 10.3
Form @ 25C: Liquid Pour Pt. 19
Viscosity cSt 100F: 64

TRYDET 2682 Ethoxylated Mixed Rosin and Fatty Acids is a relatively low-foaming detergent and emulsifier that performs exceptionally well with alkyl aryl sulfonates and soaps. Detergency is improved by incorporating conventional phosphate builders. It is used in the textile industry as a leveling agent for package and skein dyeing and in soaping off vat and naphthol dyeings. It is a versatile and strong co-emulsifier for xylene, kerosene, trichlorobenzene, o-dichlorobenzene, and similar solvents when used with TRYLON 6702 or TRYFAC 5553.

HLB: 13.4
Form @ 25C: Liquid Pour Pt. 17
Viscosity cSt 100F: 209

TRYDET 2681 Ethoxylated Tall Oil is used in formulating controlled foaming detergents for commercial laundries, textile scouring, metal and hard surface cleaners, and degreasers.

HLB: 12.1
Form @ 25C: Liquid Pour Pt. 20
Viscosity cSt 100F: 196

QUANTUM CHEMICAL CORP.: TRYCOL Ethoxylated Alcohols:

The TRYCOL Ethoxylated Alcohols include products that are excellent detergents, wetting agents, and efficient emulsifiers, used as dispersants, solubilizers, coupling agents, and rewetting agents. They provide detergency and wetting in specialty cleaners for the institutional, industrial, household, and textile markets. They are also used as textile dyeing assistants, and are versatile emulsifiers for use in industrial, textile, and consumer specialty products. The strong penetrating properties of these products suggest their selective use in agricultural chemicals.

TRYCOL 5950 POE (4) Decyl Alcohol is a high speed wetting agent with moderately low foaming properties. It is used as a penetrant for yarn and other fabrics in atmospheric and pressure dyeing systems, and in resin pad-bath applications. TRYCOL 5950 is also an intermediate in the synthesis of anionic surfactants.

HLB: 10.5
Form @ 25C: Liquid Pour Pt. -5
Viscosity cSt 100F: 17

TRYCOL 5951 POE (5) Decyl Alcohol is a wetting agent that permits better penetration into clay soils and which also acts as an efficient dispersing agent.

HLB: 11.6
Form @ 25C: Liquid Pour Pt. 8
Viscosity cSt 100F: 19

TRYCOL 5952 POE (6) Decyl Alcohol is a nonionic, high speed wetting agent which improves the permeability of clay soils promoting rapid water penetration. It is also used as a "wetter" for fire fighting, particularly of forest fires. TRYCOL 5952 provides low re-wetting properties making it useful as a pad-bath penetrant in water and grease repellent applications. It is also a good emulsifier for polyethylene emulsions used in water repellent applications

HLB: 12.4
Form @ 25C: Liquid Pour Pt. 8
Viscosity cSt 100F: 26

TRYCOL 5953 POE (6) Decyl Alcohol, 90% active is similar in most properties to TRYCOL 5952 but contains 10% water so as to effect a clear, pourable product.

HLB: 12.4
Form @ 25C: Liquid Pour Pt. -15
Viscosity cSt 100F: 28

QUANTUM CHEMICAL CORP.: TRYCOL Ethoxylated Alcohols(Continued):

TRYCOL 5993 POE (3) Tridecyl Alcohol, an oil-soluble liquid emulsifier, provides anti-foaming properties in textile formulations. It is also used as an intermediate in the manufacture of anionic surfactants by sulfation or phosphation.

HLB: 7.9
Form @ 25C: Liquid Pour Pt. -15
Viscosity cSt 100F: 19

TRYCOL 5940 POE (6) Tridecyl Alcohol has outstanding emulsifying, dispersing and wetting properties. Being a moderately low foamer, it is useful as an emulsifier and detergent in degreasers and cutting oils. It is also excellent for low temperature wool scouring. TRYCOL 5940 may be further reacted by sulphation or phosphation to form specialty surfactants.

HLB: 11.4
Form @ 25C: Liquid Pour Pt. 12
Viscosity cSt 100F: 32

TRYCOL 5949 POE (8) Tridecyl Alcohol is a water-soluble surfactant. It is useful as a foam builder and solubilizer for alkyl aryl sulfonates, essential oils, aromatic solvents, fats, and waxes. TRYCOL 5949 is also used as a co-emulsifier.

HLB: 12.5
Form @ 25C: Liquid Pour Pt. 8
Viscosity cSt 100F: 39

TRYCOL 5941 POE (9) Tridecyl Alcohol, a water-soluble nonionic general purpose wetting agent, re-wetting agent and detergent, is used with alkyl aryl sulfonates in light and heavy duty high foaming cleaners. It is also used as a re-wetter for paper towels, as a raw wool detergent, and a versatile co-emulsifier for aromatic and aliphatic solvents.

HLB: 13.0
Form @ 25C: Liquid Pour Pt. 20
Viscosity cSt 100F: 43

TRYCOL 5942 POE (11) Tridecyl Alcohol is a water-soluble surfactant used in light and heavy duty cleaning formulations.

HLB: 13.8
Form @ 25C: Liquid Pour Pt. 17
Viscosity cSt 100F: 47

QUANTUM CHEMICAL CORP.: TRYCOL Ethoxylated Alcohols(Continued):

TRYCOL 5968 POE (8) Tridecyl Alcohol, 90% active is similar in most properties to TRYCOL 5949, but contains 10% water for ease in handling.

HLB: 12.5
Form @ 25C: Liquid Pour Pt. <-10
Viscosity cSt 100F: 51

TRYCOL 5944 POE (9) Tridecyl Alcohol, 85% active is similar in most properties to TRYCOL 5941, but contains 15% water for ease in handling.

HLB: 13.0
Form @ 25C: Liquid Pour Pt. <-10
Viscosity cSt 100F: 61

TRYCOL 5943 POE (12) Tridecyl Alcohol is a hydrophilic general purpose nonionic with largely the same end-uses as TRYCOL 5942, but with a slightly higher cloud point.

HLB: 14.5
Form @ 25C: Semi-solid M.P. 16

TRYCOL 5874 POE (14) Tridecyl Alcohol, 75% active is a hydrophilic, general purpose emulsifier, containing 25% water for ease of handling.

HLB: 15.0
Form @ 25C: Liquid Pour Pt. 10
Viscosity cSt 100F: 78

TRYCOL 5946 POE (18) Tridecyl Alcohol, a strongly hydrophilic emulsifier, dispersant, solubilizer and detergent, is used as an emulsifier in cleaners and as an intermediate in the manufacture of anionic surfactants. In textile applications, TRYCOL 5946 is used as an acid dye leveling agent and as a wet processing detergent exhibiting foam stabilizing properties.

HLB: 16.0
Form @ 25C: Solid M.P. 38

QUANTUM CHEMICAL CORP.: TRYCOL Ethoxylated Alcohols(Continued):

TRYCOL 5882 POE (4) Lauryl Alcohol functions as a co-emulsifier for silicone in cleaner polishes and mold release agents, and as an all purpose oil and fat emulsifier in industrial lubricants. For textile applications, this biodegradable, oil-soluble, water-dispersible ether is used as an emulsifier for mineral oil in lubricants such as coning oils. When sulfated, it forms a high-foaming anionic surfactant.

HLB: 9.2
Form @ 25C: Liquid Pour Pt. 12
Viscosity cSt 100F: 20

TRYCOL 5963 POE (8) Lauryl Alcohol is a water-soluble, biodegradable detergent, wetting agent and emulsifier. It is a general purpose nonionic surfactant with many diverse applications.

HLB: 12.6
Form @ 25C: Liquid Pour Pt. 25
Viscosity cSt 100F: 35

TRYCOL 5967 POE (12) Lauryl Alcohol serves as a nonionic all-purpose detergent and emulsifier. It is similar to TRYCOL 5963 in diversity, but is more hydrophilic. TRYCOL 5967 is also highly biodegradable.

HLB: 14.4
Form @ 25C: Solid M.P. 32

TRYCOL 5964 POE (23) Lauryl Alcohol is a water-soluble emulsifier and solubilizing agent. It is used as a co-emulsifier for silicone in cleaner polishes and mold release agents, and as a solvent emulsifier for textile dye carriers.

HLB: 16.7
Form @ 25C: Solid M.P. 40

TRYCOL 5888 POE (20) Stearyl Alcohol is a water-soluble emulsifier and solubilizing agent. It is used as a component in dyeing assistants and as a low-foaming solvent emulsifier in textile dye carriers. It acts as a stabilizer in natural and synthetic latices, and is used as a wax emulsifier in coatings for citrus fruits.

HLB: 15.3
Form @ 25C: Solid M.P. 40

QUANTUM CHEMICAL CORP.: TRYCOL Ethoxylated Alcohols(Continued):

TRYCOL 5972 POE (23) Oleyl Alcohol is a strongly hydrophilic emulsifier, dispersant, solubilizer and detergent. It is used as a stabilizer and anticoagulant for natural and synthetic latices and dye pastes, as a dyeing assistant for wool/acrylic blends, and as a detergent and lubricant for fiber/fabric scouring. TRYCOL 5972 is also used as an emulsifier for waxes used in coating citrus fruits.

HLB: 15.8
Form @ 25C: Solid M.P. 47

TRYCOL 5971 POE (20) Oleyl Alcohol is a high hydrophilic emulsifier, dispersant and solubilizer.

HLB: 15.3
Form @ 25C: Solid M.P. 39

TRYCOL 5966 Ethoxylated Lauryl Alcohol is used as a general purpose, low foaming emulsifier, and is especially useful for mineral oils and dye carriers.

HLB: 8.7
Form @ 25C: Liquid Pour Pt. -7
Viscosity cSt 100F: 23

QUANTUM CHEMICAL CORP.: TRYCOL Ethoxylated Alkylphenols:

The TRYCOL Ethoxylated Alkylphenols have chemical and physical properties similar to the TRYCOL Ethoxylated Alcohols. They are used as dispersants, solubilizers, coupling agents, wetting agents, leveling agents and dyeing assistants.

TRYCOL 6975 POE (30) Octylphenol, 70% is an aqueous solution of a POE (30) Octylphenol that is used as an emulsifier and wetting agent.

HLB: 17.1
Form @ 25C: Liquid Pour Pt. 5
Viscosity cSt 100F: 260

TRYCOL 6984 POE (40) Octylphenol, 70% is an aqueous solution of a POE (40) Octylphenol. A strongly hydrophilic dispersing and wetting agent, its major application is as a primary emulsifier in emulsion polymerization of acrylic and vinyl monomers.

HLB: 17.9
Form @ 25C: Liquid Pour Pt. 13
Viscosity cSt 100F: 220

TRYCOL 6960 POE (1) Nonylphenol, an oil-soluble surfactant and co-emulsifier, is used in combination with water-soluble surfactants as a defoaming agent.

HLB: 4.6
Form @ 25C: Liquid Pour Pt. -10
Viscosity cSt 100F: 150

TRYCOL 6961 POE (4) Nonylphenol is an oil-soluble co-emulsifier often used as a corrosion inhibitor in two cycle engine oils.

HLB: 8.9
Form @ 25C: Liquid Pour Pt. -10
Viscosity cSt 100F: 472

TRYCOL 6940 POE (5) Nonylphenol is an oil soluble emulsifier.

HLB: 9.9
Form @ 25C: Liquid Pour Pt. 11
Viscosity cSt 100F: 102

QUANTUM CHEMICAL CORP.: TRYCOL Ethoxylated Alkylphenols
(Continued):

TRYCOL 6962 POE (6) Nonylphenol is used as a dispersing
agent, wetting agent or co-emulsifier in acid cleaning solutions,
solvent emulsions, and detergents.

HLB: 10.9
Form @ 25C: Liquid Pour Pt. -10
Viscosity cSt 100F: 100

TRYCOL 6963 POE (7) Nonylphenol is an emulsifier, dispersing
and wetting agent.

HLB: 11.7
Form @ 25C: Liquid Pour Pt. -10
Viscosity cSt 100F: 100

TRYCOL 6964 POE (9) Nonylphenol is a water-soluble surfactant
widely used for general purpose detergency, wetting, and emulsification.

HLB: 13.0
Form @ 25C: Liquid Pour Pt. 5
Viscosity cSt 100F: 112

TRYCOL 6974 POE (10) Nonylphenol is a dispersant, emulsifier
and wetting agent.

HLB: 13.2
Form @ 25C: Liquid Pour Pt. 11
Viscosity cSt 100F: 111

TRYCOL 6965 POE (11) Nonylphenol is a water-soluble surfactant
used in formulations for detergents and emulsifications.

HLB: 13.5
Form @ 25C: Liquid Pour Pt. 5
Viscosity cSt 100F: 116

TRYCOL 6953 POE (12) Nonylphenol is a water-soluble surfactant
used in formulations for detergents and emulsifications.

HLB: 14.1
Form @ 25C: Liquid Pour Pt. 10
Viscosity cSt 100F: 131

QUANTUM CHEMICAL CORP.: TRYCOL Ethoxylated Alkylphenols
(Continued):

TRYCOL 6958 POE (13) Nonylphenol is a water soluble surfactant used as a detergent, wetting agent and emulsifier.

HLB: 14.4
Form @ 25C: Liquid Pour Pt. 10
Viscosity cSt 100F: 126

TRYCOL 6952 POE (15) Nonylphenol is a moderately high-foaming, water soluble surfactant used as a wetting agent and emulsifier.

HLB: 15.0
Form @ 25C: Liquid Pour Pt. 20
Viscosity cSt 100F: 139

TRYCOL 6967 POE (20) Nonylphenol is a moderately high foaming water-soluble surfactant used as a detergent, wetting agent and emulsifier for emulsion polymerization.

HLB: 16.0
Form @ 25C: Solid M.P. 34

TRYCOL 6968 POE (30) Nonylphenol is a water-soluble detergent and wetting agent. It is also used as an emulsifier in emulsion polymerization and as an emulsifying agent for fats, oils, and waxes.

HLB: 17.1
Form: Solid M.P. 43

TRYCOL 6969 POE (30) Nonylphenol, 70% is an aqueous solution of TRYCOL 6968. Its properties and applications are generally the same, but the liquid form makes it more practical for some applications.

HLB: 17.1
Form @ 25C: Liquid Pour Pt. 5
Viscosity cSt 100F: 260

TRYCOL 6957 POE (40) Nonylphenol is a water-soluble detergent and wetting agent especially effective for high-temperature detergency and dispersing. It is a co-emulsifier for fats, waxes, and oils, as well as an effective wetting agent in solutions of electrolytes. It can be used as a stabilizer for synthetic fabrics.

HLB: 17.8
Form @ 25C: Solid M.P. 40

QUANTUM CHEMICAL CORP.: TRYCOL Ethoxylated Alkylphenols (Continued)

TRYCOL 6970 POE (40) Nonylphenol, 70% is an aqueous solution of TRYCOL 6957. Its properties and applications are generally the same, but the liquid form makes it more practical for some applications.

    HLB: 17.8
    Form @ 25C: Liquid Pour Pt. 7
    Viscosity cSt 100F: 385

TRYCOL 6971 POE (50) Nonylphenol performs similarly to TRYCOL 6957, and is slightly more hydrophilic in nature.

    HLB: 18.2
    Form @ 25C: Solid M.P. 54

TRYCOL 6972 POE (50) Nonylphenol is a 70% solution of TRYCOL 6971. Its properties and applications are generally the same, but the liquid form makes it more practical for some applications.

    HLB: 18.2
    Form @ 25C: Liquid Pour Pt. -4
    Viscosity cSt 100F: 440

TRYCOL 6942 POE (100) Nonylphenol, is a highly water-soluble, strongly hydrophilic surfactant used as a wetting agent in high electrolyte solutions and as a stabilizer in synthetic latices.

    HLB: 19.0
    Form @ 25C: Solid M.P. 56

TRYCOL 6981 POE (100) Nonylphenol is a 70% solution of TRYCOL 6942. Its properties and applications are generally the same, but the liquid form makes it more practical for some applications.

    HLB: 19.0
    Form @ 25C: Liquid Pour Pt. 18
    Viscosity cSt 100F: 564

TRYCOL 6954 POE (150) Nonylphenol is a highly water soluble, strongly hydrophilic surfactant used as a wetting agent and dispersant.

    HLB: 19.3
    Form @ 25C: Solid M.P. 60

QUANTUM CHEMICAL CORP.: TRYCOL Ethoxylated Alkylphenols (Continued):

TRYCOL 6985 POE (8) Dinonylphenol, a moderately lipophilic emulsifier and foam control agent, is useful as an emulsifier for polar and non-polar solvents, as a spreading agent in pigment printing, and as a post-stabilizer in emulsion polymerization. TRYCOL 6985 is also used as an intermediate in the synthesis of anionic surfactants, as an emulsifier in textile jet dye carrier applications and as a surfactant component in acidic cleaners, aerosols, insecticides and wax emulsions.

HLB: 10.4
Form @ 25C: Liquid Pour Pt. 9
Viscosity cSt 100F: 173

TRYCOL 6989 POE (150) Dinonylphenol, 50% is an aqueous solution of POE (150) dinonylphenol. It is a strongly hydrophilic emulsifier, dispersing and wetting agent. Its liquid form makes it more practical in applications such as controlled suds component in built, heavy duty detergents, hard surface cleaners, alkaline bottle washing compounds and dairy detergents. It may also be used as a dispersing and wetting agent for pesticides. Textile uses include an emulsifier for flaked biphenyl dye carriers, a detergent component and leveling agent for acid, cationic and premetalized dyeings of wool and acrylic fibers.

HLB: 19.0
Form @ 25C: Liquid Pour Pt. 23
Viscosity cSt 100F: 367

QUANTUM CHEMICAL CORP.: TRYFAC Phosphate Esters:

TRYFAC Phosphate Esters are available as free acids and as various salts. The salts of these phosphate esters, made by neutralizing the free acid form with different alkalies are used as emulsifiers, wetting agents, detergents, antistats, metal lubricants, corrosion inhibitors, agricultural adjuvants and hydrotropes.

TRYFAC 5559 Phosphate Ester is a water soluble detergent and wetting agent. It is a good detergent and oily soil emulsifier in built liquid detergents. TRYFAC 5559 is used as an antistat in fiber finishes for polyester and polypropylene.

Form @ 25C: Liquid Pour Pt. 18
Viscosity cSt 100F: 560

TRYFAC 5552 Phosphate Ester (free acid form) is an anionic surfactant intermediate. End-use properties depend on the salt formed when the product is neutralized. Various salts are excellent emulsifiers for aliphatic and aromatic solvents.

Form @ 25C: Liquid Pour Pt. < -15
Viscosity cSt 100F: 170

TRYFAC 5553 Phosphate Ester (potassium salt of TRYFAC 5552) is an anionic, water-soluble emulsifier, wetting agent, detergent and antistat. Industrially, it is used as an emulsifier for chlorinated hydrocarbons in heavy duty cleaners and alkalitolerant detergents, and as a component in metalworking compounds. In the textile industry it is used as an emulsifier and detergent in solvent scouring, as an emulsifier for dye carriers, as a scouring agent for cotton goods, and as an antistat in processing oils for natural and synthetic fibers. TRYFAC 5553 also provides corrosion inhibition and outstanding dispersion properties.

Form @ 25C: Liquid Pour Pt. <-15
Viscosity cSt 100F: 345

TRYFAC 5554 Phosphate Ester (potassium salt) is similar to TRYFAC 5553 in properties and functions, but is more water soluble. In addition to its excellent surface active properties, it can also be used as a polymer emulsification stabilizer. In the conversion of starches in textile desizing, TRYFAC 5554 acts as a detergent and wetting agent, and does not interfere with enzymatic action.

Form @ 25C: Liquid Pour Pt. -9
Viscosity cSt 100F: 340

Emulsifying Agents

QUANTUM CHEMICAL CORP.: TRYFAC Phosphate Esters(Continued):

TRYFAC 5555 Complex Phosphate Ester (free acid form) is the most hydrophobic ester of the series. In the textile industry it is used as an emulsifier in low foaming dye carrier systems, and in solvent scouring systems. It is also an excellent emulsifier for various flame retardants. The stability of TRYFAC 5555 allows it to be used in high-temperature applications.

Form @ 25C: Liquid Pour Pt. -3
Viscosity cSt 100F: 2300

TRYFAC 5556 Complex Phosphate Ester (free acid form) is a wetting agent, dispersant and antistat in textile processing. It is a good solvent emulsifier for textile scours, detergents, and for pesticide formulations. It is stable in alkaline built liquid detergent formulations and tolerates salts. It is a dry-cleaning detergent and is also used in emulsion polymerization.

Form @ 25C: Liquid Pour Pt. 5
Viscosity cSt 100F: 1700

TRYFAC 5569 Phosphate Ester is a 100% active product in free acid form. It does not discolor in contact with solid caustic. A five percent solution of TRYFAC 5569 in 10% aqueous caustic is clear and low foaming. TRYFAC 5569 acts as a hydrotrope toward nonionic surfactants in alkaline solution.

Form @ 25C: Liquid Pour Pt. 5
Viscosity cSt 100F: 1180

TRYFAC 5571 Phosphate Ester is the potassium salt of a phosphate ester of an ethoxylated alcohol. It is a detergent, dispersant, and wetting agent stable in dilute caustic solutions. Among its uses are fabric preparation, hard surface cleaning, and antistatic protection of fibers.

Form @ 25C: Liquid Pour Pt. <-15
Viscosity cSt 100F: 650

TRYFAC 5573 Phosphate Ester is an oleophilic ester in free acid form. It is used as a mold release agent, an antistat, and as a dispersant/emulsifier. TRYFAC 5573 can be neutralized with a variety of organic and inorganic bases.

Form @ 25C: Solid M.P. 35

QUANTUM CHEMICAL CORP.: TRYFAC Phosphate Esters(Continued):

TRYFAC 5576 Phosphate Ester is the potassium salt of a phosphate ester of an ethoxylated alcohol. It is very hydrophilic, featuring good tolerance to electrolytes and caustic solutions. TRYFAC 5576 is a moderately foaming surfactant in dilute caustic. It has good antistatic properties, is an anticorrosive and dispersant.

Form @ 25C: Liquid Pour Pt. <-10
Viscosity cSt 100F: 190

TRYFAC 5557 Phosphate Ester is the potassium salt of a phosphate ester. It is a 55% active surfactant that is stable in highly alkaline solutions. TRYFAC 5557 is a rapid wetter and dispersant, especially useful in formulating textile scours.

Form @ 25C: Liquid Pour Pt. 5

TRYFAC 5560 Phosphate Ester (free acid form) can be used as an emulsifier, dispersant and antistat. TRYFAC 5560 is listed under 21 CFR 178.3400.

Form @ 25C: Liquid Pour Pt. 5

QUANTUM CHEMICAL CO.: TRYLOX Ethoxylated Sorbitol and Ethoxylated Sorbitol Esters:

TRYLOX Ethoxylated Sorbitols and Ethoxylated Sorbitol Esters are hydrophilic surfactants that function as emulsifiers, dispersants, wetting agents, lubricants, plasticizers, and solubilizers in household, industrial, and textile specialty products. The TRYLOX series generally contains structures that are highly efficient oil/solvent emulsifiers, hence they find wide acceptance as major emulsifier components in industrial and textile concentrates.

Product Name:

TRYLOX 6753 POE (20) Sorbitol functions as a humectant and plasticizer. It serves as an intermediate in the synthesis of fatty acid esters and improves the pourability and clarity of high solid "clear concentrate" surfactant solutions.

HLB: 15.4
Form @ 25C: Liquid Pour Pt. 7
Viscosity cSt 100F: 200

TRYLOX 6746 POE (40) Sorbitol Hexaoleate is an O/W balanced emulsifier for petroleum oils, vegetable oils, and solvents. It functions as an emulsifier for oil based metal lubricants and textile processing aids and is a superior emulsifier component for aliphatic and aromatic solvents in low foaming textile dye carriers.

HLB: 10.4
Form @ 25C: Liquid Pour Pt. <-10
Viscosity cSt 100F: 120

TRYLOX 6747 POE (60) Sorbitol Hexaoleate, a low foaming emulsifier for organic solvents, can also be used as an emulsifier for paraffin oils in industrial lubricants. Its balanced solubility properties make it useful in various systems and in some instances can be used as the primary emulsifier.

HLB: 11.3
Form @ 25C: Liquid Pour Pt. 4
Viscosity cSt 100F: 110

QUANTUM CHEMICAL CORP.: TRYLOX Ethoxylated Triglycerides:

The TRYLOX Ethoxylated Triglycerides are versatile nonionic surfactants which range in water solubility from lipophilic to strongly hydrophilic. They are used as emulsifiers, dispersants, rewetting agents, solubilizers, lubricants, softeners, antistatic agents, latex stabilizers, leather processing auxiliaries, and dyeing assistants for textiles, paper and leather.

TRYLOX 5900 POE (5) Castor Oil, a hydrophobic emulsifier and dispersing agent, is an excellent co-emulsifier exhibiting foam control properties for chlorinated and aromatic solvents. In water-based paints, it aids the dispersion of pigment slurries and improves gloss and freeze-thaw resistance. TRYLOX 5900 is also a clay dispersant and carrier for paper coatings. Textile uses include foam control and emulsification in low foaming dye carriers.
HLB: 4.0
Form @ 25C: Liquid Pour Pt. -3
Viscosity cSt 100F: 375

TRYLOX 5902 POE (16) Castor Oil is a moderately lipophilic emulsifier and lubricant. It is used as a co-emulsifier for metalworking oils and hydraulic fluids. In textiles it is used as a component in leveling and dispersing agents for vat and naphthol dyes, and as a co-emulsifier for rayon delustrants and fiber lubricants.
HLB: 8.6
Form @ 25C: Liquid Pour Pt. -22
Viscosity cSt 100F: 546

TRYLOX 5904 POE (25) Castor Oil is a liquid water-soluble emulsifier and lubricant. It is recommended for the formulation of soluble oils, cutting fluids, and fiber finishes.
HLB: 10.8
Form @ 25C: Liquid Pour Pt. -5
Viscosity cSt 100F: 396

TRYLOX 5906 POE (30) Castor Oil is a water-soluble emulsifier for oils, solvents, and waxes, and a dispersant for pigments. It is used as a degreaser, emulsifier, and lubricant in fat liquoring formulas, as a viscosity and emulsion stabilizer for polyvinyl acetate and water-based paints, and as an emulsifier and dispersing agent for urethane foams and polyester resins. TRYLOX 5906 contributes to the softness and absorbency of wet-strength papers. In textile applications it is used as a co-emulsifier in fabric softener and dye carrier systems, as a dyeing assistant, and as an emulsifier in synthetic fiber lubricants.
HLB: 11.8
Form @ 25C: Liquid Pour Pt. 9
Viscosity cSt 100F: 309

QUANTUM CHEMICAL CORP.: TRYLOX Ethoxylated Triglycerides (Continued):

TRYLOX 5907 POE (36) Castor Oil is a water-soluble emulsifier for solvents and oils. It is a lubricant and softener for textiles and leather. It also acts as a versatile co-emulsifier with anionic emulsifiers in formulating textile dye carriers based on aromatic and aliphatic solvents.

HLB: 12.6
Form @ 25C: Liquid Pour Pt. 12
Viscosity cSt 100F: 363

TRYLOX 5909 POE (40) Castor Oil is more water-soluble than TRYLOX 5907, but generally has similar applications.

HLB: 13.0
Form @ 25C: Liquid Pour Pt. 18
Viscosity cSt 100F: 313

TRYLOX 5918 POE (200) Castor Oil, 50% a strongly hydrophilic emulsifier, lubricant and antistat, is used in textile fiber processing as an antistatic, humectant and scrooping agent. It is also used as an emulsifier for hydrophobic glycerides and sorbitan esters in blended lubricants for fibers and yarns.

HLB: 18.1
Form: Liquid Pour Pt. 7
Viscosity cSt 100F: 1015

TRYLOX 5921 POE (16) Hydrogenated Castor Oil is an emulsifier, lubricant and softener having very low odor and excellent heat stability. It is an excellent emulsifier for castor oil and provides lubricity and softness in fabric softeners and aerosol fabric sprays.

HLB: 8.6
Form @ 25C: Liquid Pour Pt. 7
Viscosity cSt 100F: 569

TRYLOX 5922 POE (25) Hydrogenated Castor Oil is a moderately hydrophilic emulsifier used in textile applications as an emulsifier for resin finishing and softener-lubricant systems. It is a good co-emulsifier in dye carrier solvent systems and for lanolin.

HLB: 10.8
Form @ 25C: Liquid Pour Pt. 5
Viscosity cSt 100F: 535

QUANTUM CHEMICAL CORP.: TRYMEEN Ethoxylated Fatty Amines:

The TRYMEEN Ethoxylated Fatty Amines are mildly cationic surfactants. They are used as wetting and penetrating agents, and are substantive to a wide variety of substrates, e.g., metals, glass, textiles, plastics and clays. When the TRYMEEN cationics are absorbed, they deposit emulsified oils and/or dispersed solids on the surfaces upon which they are exhausted, usually without breaking the emulsions or dispersions. The products are also effective emulsifiers, dispersants, solubilizers, antistats, lubricants and corrosion inhibitors.

TRYMEEN 6603 POE (8) Tallow Amine is a moderately low foaming emulsifier for mineral oil, waxes, solvents, and vegetable oils and contributes to the wetting, lubricating, and softening properties of the end product. It is used as a solvent emulsifier for low-foaming dye carriers, as a dyeing assistant, and as a cationic emulsifier for polyethylene textile softeners. TRYMEEN 6603 is also an emulsifier for fats and oils in industrial lubricants.

HLB: 11.4
Form @ 25C: Liquid Pour Pt. -2
Viscosity cSt 100F: 87

TRYMEEN 6606 POE (15) Tallow Amine, a hydrophilic emulsifier, is a highly effective anti-precipitant for mixed dye baths, a superior leveling agent for acid dyes, and a migrating agent for dispersed dyes. TRYMEEN 6606 is also an intermediate for quaternary ammonium compounds and a versatile antistat for processing synthetic fibers.

HLB: 14.3
Form @ 25C: Liquid Pour Pt. -10
Viscosity cSt 100F: 96

TRYMEEN 6607 POE (20) Tallow Amine is used in the textile industry as an antistat and lubricant for wool and synthetic fiber processing, and as a co-emulsifier and antistat in synthetic fiber spin finishes. It is also an anti-precipitant, leveling, and migrating agent in various dyeing procedures, and an antistat in carpet shampoos.

HLB: 15.4
Form @ 25C: Liquid Pour Pt. -2
Viscosity cSt 100F: 119

QUANTUM CHEMICAL CORP.: TRYMEEN Ethoxylated Fatty Amines (Continued):

TRYMEEN 6609 POE (25) Tallow Amine has properties and uses similar to TRYMEEN 6607 but is somewhat more hydrophilic.
HLB: 16.0
Form @ 25C: Liquid Pour Pt. 16
Viscosity cSt 100F: 128

TRYMEEN 6637 POE (40) Tallow Amine is an 80% aqueous solution of a strongly hydrophilic emulsifier that is used as an antistatic additive for commercial carpet maintenance. It is also used as a dyeing assistant and as a stabilizer for natural and synthetic latices to prevent coagulation by acids.
HLB: 17.4
Form @ 25C: Liquid Pour Pt. 9
Viscosity cSt 100F: 150

TRYMEEN 6617 POE (50) Stearyl Amine is a strongly hydrophilic emulsifier and antistat used as an emulsifier in metal buffing compounds. In latex rubber compounding it is used to emulsify stearic acid and to prevent premature coagulation in mild acid or salt baths. Textile applications include use as a lubricant for fiberglass and as a leveling agent for acid, cationic and disperse dyeings.
HLB: 17.8
Form @ 25C: Solid M.P. 35

TRYMEEN 6601 POE (10) Coco Amine is used as a co-emulsifier and antistat for textile processing oils, as a dispersing agent for inorganic salts in viscose spinning, and as an emulsifier for fats and oils in industrial lubricants.
HLB: 13.6
Form @ 25C: Liquid Pour Pt. -10
Viscosity cSt 100F: 68

TRYMEEN 6623 POE (30) Oleyl Amine is a hydrophilic emulsifier and textile dyeing assistant. It functions as an antiprecipitant in cross dyeing, and as a mild stripping agent and dye leveler for acid dyes.
HLB: 16.6
Form @ 25C: Solid M.P. 35

TRYMEEN 6622 POE (30) Oleyl Amine is an 80% aqueous solution of TRYMEEN 6623.
HLB: 16.6
Form @ 25C: Liquid Pour Pt. 5
Viscosity cSt 100F: 445

TRYMEEN 6640 POE (15) Tallow Propylene Diamine is an excellent dispersant in acidic solutions. It is a very strong retarder of acidic, disperse, and premetallized dyestuffs.
HLB: 13.1
Form @ 25C: Liquid Pour Pt. <-10
Viscosity cSt 100F: 120

QUANTUM CHEMICAL CORP.: Miscellaneous Surfactants:

TRYCOL 6837 Nonionic Ether is a versatile emulsifier for solvents, vegetable oils, and waxes. In the leather industry, it is used as an emulsifier for tanning chemicals. Textile uses include resin bath penetrants, polymer stabilizers, solvent scour emulsifiers, and enzyme bath penetrants. It is also used in cold water scours for felted fabrics.

Form @ 25C: Liquid Pour Pt. 5
Viscosity cSt 100F: 126

TRYLON 6702 Anionic is used as a solvent emulsifier in degreasing cleaners for metal parts and engine blocks. In this application, it exhibits good detergency and excellent rinsability. In the textile industry, TRYLON 6702 is used as an emulsifier for dye carriers in high-temperature dye applications. It is also used as an emulsifier for solvent scouring. This hydrophilic, anionic, general-purpose emulsifier is effective for a wide range of aliphatic and aromatic solvents including stoddard solvent, kerosene, xylene, and chlorinated solvents.

Form @ 25C: Liquid Pour Pt. -10
Viscosity cSt 100F: 1250

TRYLON 6735 Nonionic Wetting Agent is a low-foaming emulsifier. It finds applications in industrial, institutional and consumer detergents. In the textile industry, it is used in scouring, wetting and penetrating operations where low foam is required. It is also a co-emulsifier for solvents in solvent and dye carriers.

Form @ 25C: Liquid Pour Pt. 9
Viscosity cSt 100F: 41

EMERY 6885 Emulsifier is a liquid, fatty-based product developed specially for the emulsification of a wide variety of triglycerides, including fats and oils.

HLB: 7.75
Form @ 25C: Liquid Pour Pt. -14
Viscosity cSt 100F: 149

SANDOZ CHEMICALS: Chemical 39 Base:

CTFA designation: Stearamidoethyl Ethanolamine

Uses in the Cosmetic and Toiletry Industry include:
- A cationic emulsifying agent for creams and lotions.
- An additive to hydro-alcoholic and alcoholic astringent solutions giving these products lubricity without oiliness.
- An additive to hair products to give a conditioning effect.
- An additive to skin products, such as shaving creams, giving added lubricity.

CHEMICAL 39 BASE is a cationic emulsifier of the fatty amidoamine classification. The hydrophobe is derived from stearic acid. The hydrophile is derived from hydroxydiamine.

Typical Properties:
  Color: Gardner 3
  Odor: Very slight amine odor
  Equivalent weight: 636
  Solubility: Insoluble in water, dispersible as the amine salt.

As a secondary emulsifier with glyceryl monostearate, CHEMICAL 39 BASE, as the amine salt, forms excellent, stable, and elegant cationic oil/water emulsions of virtually any desired viscosity. The resulting acid creams and lotions are extremely tolerant of salt and compatible with most cationic ingredients. The presence of the hydrophobic cationic emulsifier lends a smooth, velvety feel to the skin even after the finished product is rinsed off.

CHEMICAL 39 BASE is normally added to the oil phase of the emulsion, the neutralizing acid to the water phase. The amount of acid used is partly dependent on the final desired pH. The equivalent weight of 636 may be used to determine a stoichiometric amount of the appropriate acid.

SANDOZ CHEMICALS: CHEMICAL BASE 6532:

CTFA designation: Stearamidoethyl Diethylamine

Uses in the Cosmetic and Toiletry Industry include:
- A cationic emulsifying agent for creams and lotions.
- An additive to hydro-alcoholic and alcoholic astringent solutions giving these products lubricity without oiliness.
- An additive to hair products to give a conditioning effect.
- An additive to skin products, such as shaving creams, giving added lubricity.

CHEMICAL BASE 6532, as supplied, is not soluble in water but becomes water soluble when neutralized with an acid. Once the water soluble amine-salt is made(pH of 4-6, depending upon the acid used), the pH can subsequently be raised to 6.5-7.5 with no loss in solubility.

Specifications:
    Structure: N,N-Diethyl-N'-Stearoyl-Ethylene-Diamine or Stearamidoethyl Diethylamine
    Appearance: Cream to pale yellowish-tan, hard, wax-like crystalline solid.
    Solidification Point: 49C Minimum
    Equivalent Weight by Titration: 362-382
                                       Theoretical = 382
    Acid Number mg KOH/g: 3 maximum
    Assay: 97% minimum

Preparation of Water-Soluble Salts of CHEMICAL BASE 6532:

When formulating an oil-in-water cream or lotion, the salt of CHEMICAL BASE 6532 can be prepared via two methods:

1) Dissolve the CHEMICAL BASE 6532 in the oil phase. Add the preferred acid to the water phase. Add oil phase to the water phase.

2) Prepare a premix of CHEMICAL BASE 6532 and the preferred acid. Add the premix to either the oil or water phase of a cream or lotion formula.

SANDOZ CHEMICALS: SANDOPAN Carboxylated Surfactants:

Trade Name:

SANDOPAN DTC:
  CTFA Name: Sodium Tridecth-7-Carboxylate
  Physical Form: Light yellow gel
  % Solids: 75+-2%
  Recommended Applications: Detergents, emulsifiers, wetting agents, solubilizer, cationic compatible

SANDOPAN DTC-100:
  CTFA Name: Same as above
  Physical Form: Clear yellow liquid
  % Solids: 70+-2%
  Recommended Applications: Detergents, emulsifiers, wetting agents, pH stable, good for solvent cleaners

SANDOPAN DTC Acid:
  CTFA Name: Trideceth-7-Carboxylic Acid
  Physical Form: Clear liquid
  % Solids: 90+-2%
  Recommended Applications: Detergents, emulsifiers, wetting agents. Free acid form, oil and solvent soluble

SANDOPAN DTC Linear P:
  CTFA Name: Sodium C12-15 Pareth-6-Carboxylate
  Physical Form: White, opaque semi-pourable gel
  % Solids: 70+-5%
  Recommended Applications: Detergents, emulsifiers, wetting agents, solubilizer, viscosity enhancer in certain systems.

SANDOPAN DTC Linear P Acid:
  CTFA Name: C12-15 Pareth-6-Carboxylic Acid
  Physical Form: Clear liquid
  % Solids: 90+-5%
  Recommended Applications: Detergents, emulsifiers, wetting agents. Good oil solubilizer and solvent systems.

SANDOPAN LS-24:
  CTFA Name: Sodium Laureth-13-Carboxylate
  Physical Form: Clear to slightly hazy gel
  % Solids: 69+-2%
  Recommended Applications: Mild detergent, emulsifiers, solubilizers. Good for baby shampoos and personal care products.

SANDOZ CHEMICALS: SANDOPAN Carboxylated Surfactants(Continued):

Trade Name:

SANDOPAN JA-36:
 CTFA Name: Trideceth-19-Carboxylic Acid
 Physical Form: Clear to slightly hazy liquid
 % Solids: 90+-2%
 Recommended Applications: Moderate foaming mild surfactant. Oil solubilizer, wetting agent.

SANDOPAN RS-8:
 CTFA Name: Sodium C16-20 Ethoxylate Carboxylate
 Physical Form: Off white, firm paste
 % Solids: 70+-2%
 Recommended Applications: Emulsifier, solubilizer. Water dispersible.

SANDOPAN MA-18:
 CTFA Name: Alkylaryl Ethoxylate Carboxylic Acid
 Physical Form: Clear liquid
 % Solids: 90+-3%
 Recommended Applications: Detergent, wetting agent, solubilizer, oil soluble surfactant

SANDOPAN KST:
 CTFA Name: Sodium Ceteth-13-Carboxylate
 Physical Form: Solid
 % Solids: 97+-2%
 Recommended Applications: Emulsifier, detergent, lime soap dispersant. Good for stick and soap preparations.

SANDOPAN B Liquid:
 CTFA Name: Carboxylated, C4 Paraffinic Ethoxylate
 Physical Form: Clear light amber liquid
 % Solids: 40+-5%
 Recommended Applications: Non-foaming industrial surfactant with caustic stability

SCANROAD INC.: CATIMULS Emulsifiers:

CATIMULS 220: Emulsifier for slow setting cationic bitumin emulsions:

Application:
CATIMULS 220 is used in making CSS cationic emulsions suitable for all fine graded cold-mix designs, such as: slurry seals, dense-graded mixes, sand emulsion mixes, soil stabilizations, gravel emulsion mixes, etc.
CATIMULS 220 allows coating of a wide variety of aggregate types including: limestone, volcanic materials, laterite materials, sands, granite, silicious materials etc.
CATIMULS 220 emulsions are especially suited for emulsified asphalt mixes utilizing a high fines content and or very active fines. The product is especially suited for applications in warm to very hot climates which can be either humid or dry.
CATIMULS 220 is an excellent emulsifier for use with dirty aggregates.

Physical Properties:
 Visual appearance at 77F: viscous liquid
 Flash point, open cup: above 200F
 Boiling point: 220F

CATIMULS 301 (Modified) Cationic Asphalt Emulsifier:

Application:
CATIMULS 301 (Modified) is a liquid, cationic asphalt emulsifier formulated for use in rapid and medium setting emulsions. CATIMULS 301 (Modified) can be used with a wide variety of asphalt cements.
Before use, the emulsifier must be neutralized with an acid, normally hydrochloric acid. In the preparation of the soap solution, it is necessary to first acidify the heated water (110-130F) with hydrochloric acid before adding the emulsifier.

Physical Properties:
 Visual appearance: Dark brown liquid
 Flash point, F (Pensky-Martens closed cup): above 100
 Density at 77F, lbs/gal: 7.4
 Viscosity, 77F, SFS: 75-100
 Viscosity, 100F, SFS: 15-20

SCANROAD INC.: CATIMULS Emulsifiers(Continued):

CATIMULS 305 Cationic Asphalt Emulsifier:

Application:
   CATIMULS 305 is a liquid, cationic asphalt emulsifier formulated for use in rapid and medium setting emulsions. CATIMULS 305 can be used with a wide variety of asphalt cements.
   Before use, the emulsifier must be neutralized with an acid, normally hydrochloric acid. In the preparation of the soap solution, it is necessary to first acidify the heated water (110-130F) with hydrochloric acid before adding the emulsifier. A pH of 2 to 3 is recommended for the soap solution of CRS emulsions, while a pH of 3 to 4 is suggested for the soap solution of CMS emulsions.

Physical Properties:
   Visual appearance: Dark brown liquid
   Flash point, F (Pensky-Martens closed cup): above 200
   Density at 77F, lbs/gal: 7.6+-0.1
   Viscosity, 77F, SFS: 85-110
   Viscosity, 100F, SFS: 35-45

CATIMULS 317 Cationic Asphalt Emulsifier:

Application:
   CATIMULS 317 is a liquid, cationic asphalt emulsifier designed for use as a viscosity limiter in CRS-1 and CRS-2 emulsions.

Physical Properties:
   Visual appearance: Dark brown viscous liquid
   Flash Point, F: above 200
   Density at 77F, lbs/gal: 8.25+-0.1
   Viscosity, 77F, SFS: 650-900
   Viscosity, 100F, SFS: 200-260

CATIMULS 404 Cationic Asphalt Emulsifier for Slurry Seal
   Application:

Application:
   CATIMULS 404 is a combined liquid, cationic asphalt emulsifier and additive formulated for use in modified bitumen emulsions for improved dense graded asphalt mixes, such as polymer modified slurry seals.

Physical Properties:
   Visual appearance: Brown liquid
   Flash Point, F (Pensky-Martens closed cup): above 200
   Density at 77F, lbs/gal: 7.9
   Pour Point, F: 20
   Viscosity, 77F, SFS: 225

SHELL CHEMICAL CO.: NEODOLS:

NEODOL linear primary alcohols are manufactured in various blends in the C9 to C15 carbon number range. At room temperature, they are high purity, colorless liquids or pastes and resemble fatty alcohols in chemical behavior.

NEODOL alcohols can be used in a wide variety of end uses including derivative applications requiring detergency and foaming. They can be superior in some cases to corresponding Ziegler or natural fatty alcohol analogs.

NEODOL ethoxylate nonionic surfactants consist of a variety of products which vary in both alcohol and ethylene oxide chain lengths, and provide a wide range of physical and surfactant properties. NEODOL ethoxylates are colorless and range from liquids to low melting solids of pasty consistency. They are excellent wetting agents, emulsifiers and detergents, and are moderate foamers.

NEODOL Alcohols:

91:
   Carbon chains present: C9/C10/C11
   Molecular weight: 160
   Active content, %w: 100
   Melting range, F: 3-25

1:
   Carbon chains present: C11
   Molecular weight: 173
   Active content, %w: 100
   Melting range, F: 42-57

23:
   Carbon chains present: C12/C13
   Molecular weight: 194
   Active content, %w: 100
   Melting range, F: 45-72

25:
   Carbon chains present: C12/C13/C14/C15
   Molecular weight: 203
   Active content, %w: 100
   Melting range, F: 54-77

45:
   Carbon chains present: C14/C15
   Molecular weight: 218
   Active content, %w: 100
   Melting range, F: 59-97

SHELL CHEMICAL CO.: NEODOL Ethoxylates:

Product:

91-2.5:
    EO groups/alcohol, mole/mole, avg.: 2.7
    Molecular weight: 281
    Active content, %w: 100
    EO content, %w: 42.3
    Melting range, F: -31 to -2

91-6:
    EO groups/alcohol, mole/mole, avg.: 6.1
    Molecular weight: 428
    Active content, %w: 100
    EO content, %w: 62.7
    Melting range, F: 21-52

91-8:
    EO groups/alcohol, mole/mole, avg.: 8.2
    Molecular weight: 519
    Active content, %w: 100
    EO content, %w: 69.5
    Melting range, F: 45-68

23-1:
    EO groups/alcohol, mole/mole, avg.: 1.0
    Molecular weight: 238
    Active content, %w: 100
    EO content, %w: 18.5
    Melting range, F: 27-48

23-3:
    EO groups/alcohol, mole/mole, avg.: 2.9
    Molecular weight: 322
    Active content, %w: 100
    EO content, %w: 39.6
    Melting range, F: 19-37

23-5:
    EO groups/alcohol, mole/mole, avg.: 5.0
    Molecular weight: 413
    Active content, %w: 100
    EO content, %w: 53.3
    Melting range, F: 27-61

23-6.5:
    EO groups/alcohol, mole/mole, avg.: 6.7
    Molecular weight: 488
    Active content, %w: 100
    EO content, %w: 60.4
    Melting range, F: 39-70

SHELL CHEMICAL CO.: NEODOL Ethoxylates(Continued):

Product:

23-6.5T:
    EO groups/alcohol, mole/mole, avg.: 7.6
    Molecular weight: 529
    Active content, %w: 100
    EO content, %w: 63.2
    Melting range, F: 36-66

23-12:
    EO groups/alcohol, mole/mole, avg.: 11.9
    Molecular weight: 719
    Active content, %w: 100
    EO content, %w: 72.8
    Melting range, F: 63-90

25-3:
    EO groups/alcohol, mole/mole, avg.: 3.0
    Molecular weight: 338
    Active content, %w: 100
    EO content, %w: 39.0
    Melting range, F: 27-45

25-7:
    EO groups/alcohol, mole/mole, avg.: 7.3
    Molecular weight: 524
    Active content, %w: 100
    EO content, %w: 61.3
    Melting range, F: 36-70

25-9:
    EO groups/alcohol, mole/mole, avg.: 8.9
    Molecular weight: 597
    Active content, %w: 100
    EO content, %w: 65.6
    Melting range, F: 57-77

25-12:
    EO groups/alcohol, mole/mole, avg.: 11.9
    Molecular weight: 729
    Active content, %w: 100
    EO content, %w: 71.8
    Melting range, F: 68-86

SHELL CHEMICAL CO.: NEODOL Ethoxylates(Continued):

Product:

45-2.25:
    EO groups/alcohol, mole/mole, avg.: 2.29
    Molecular weight: 319
    Active content, %w: 100
    EO content, %w: 31.6
    Melting range, F: 48-68

45-7:
    EO groups/alcohol, mole/mole, avg.: 7.1
    Molecular weight: 529
    Active content, %w: 100
    EO content, %w: 59.0
    Melting range, F: 48-75

45-7T:
    EO groups/alcohol, mole/mole, avg.: 7.9
    Molecular weight: 567
    Active content, %w: 100
    EO content, %w: 61.3
    Melting range, F: 46-73

45-13:
    EO groups/alcohol, mole/mole, avg.: 13.0
    Molecular weight: 790
    Active content, %w: 100
    EO content, %w: 72.4
    Melting range, F: 77-93

WERNER G. SMITH, INC.: Non Ionic Emulsifier T-9:

Typical Analysis:

Cloud	60-70F.
Acid No.	9-15
Sap. No.	60-70
Iodine No.	35-45
Color, Gardner	6-12
SSU @ 100F	195-250
Sp. Gr. @ 60F.	0.967-0.976
pH 1%	6.5-7.0
HLB	12

This is an excellent emulsifier for oils and emulsion systems. It is low foam. It is water dispersible. It is not a polyethylene glycol ester. It is soluble in paraffinic oils. It has a low freezing point. It is non-gelling.

It is 100% active emulsifier and contains no flash coupling agent. It is 100% biodegradable.

SONNEBORN DIVISION: BARIUM PETRONATE: Oil-Soluble Petroleum Sulfonate:

Description:
  Sonneborn barium petroleum sulfonate oil-soluble alkylaryl petroleum sulfonates are characterized by differences in total basic number.

Application:
  Basic BARIUM PETRONATE petroleum sulfonate is a natural sulfonate with an alkalinity present as barium carbonate having a Total Base Number (TBN) of 70. The product is used in additives for lube and industrial oils, fuels, specialty oils and greases.

  Neutral BARIUM PETRONATE 50-S synthetic sulfonate is free of residual alkalinity. Has application in additives for lube and industrial oils, fuel oil, undercoating and greases.

Product:

Basic:
  Barium sulfonate (%): 40-45
  Water (%), max: 0.5
  TBN (mg KOH/g): 65 min
  Molecular weight: 1100
  Color, max: 7.0
  Viscosity (SSU): 250
  Ba (%): 14.1
  Oil (%): 43.0

Neutral (50-S):
  Barium sulfonate (%): 50-52
  Water (%), max: 1.0
  TBN (mg KOH/g): 5.0 max
  Molecular weight: 1000
  Color, max: 3.5
  Viscosity (SSU): 100
  Ba (%): 6.9
  Oil (%): 49.5

SONNEBORN DIVISION: CALCIUM PETRONATE Oil-Soluble Petroleum Sulfonate:

Description:
Sonneborn calcium petroleum sulfonate oil-soluble alkylaryl natural petroleum sulfonates are characterized by differences in alkalinity and molecular weights.

Application:
PETRONATE 25H - alkalinity present as calcium hydroxide. It is used as a detergent and rust inhibitor component in lube oil additives, in rust preventive formulations and as an emulsifier for water in oil systems.
PETRONATE 25C - alkalinity present as calcium carbonate. Application similar to those of PETRONATE 25H.
PETRONATE HMW - alkalinity present as calcium hydroxide. Very high in molecular weight. Widely used as a detergent and rust inhibitor components in lubricating oil additives and in rust preventive formulations.

Product:

25H:
    Calcium sulfonate (%): 43-46
    Ca (%), min: 2.85
    Water (%), max: 1.0
    TBN (mg KOH/g), min: 21
    Molecular weight: 880

25C:
    Calcium sulfonate (%): 43-46
    Ca (%), min: 2.85
    Water (%), max: 1.0
    TBN (mg KOH/g), min: 21
    Molecular weight: 880

HMW:
    Calcium sulfonate (%): 45-46
    Ca (%), min: 2.5
    Water (%), max: 0.4
    TBN (mg KOH/g), min: 20
    Molecular weight: 974

SONNEBORN DIVISION: SODIUM PETRONATE: Oil-Soluble Petroleum Sulfonate:

Description:
Sonneborn sodium petroleum sulfonate oil-soluble alkylaryl natural petroleum sulfonates are characterized by different molecular weights.

Application:
Oil-soluble SODIUM PETRONATES are widely employed where wetting, rewetting, suspending, emulsifying and corrosion resisting properties are needed.

PETRONATE:

L:
Sulfonate (%): 61-63
Water (%): 4-5
Molecular Weight: 415-430
Color: 1.5-3R

HL:
Sulfonate (%): 61-63
Water (%): 4-5
Molecular Weight: 440-470
Color: 4-5R

CR:
Sulfonate (%): 61-63
Water (%): 4-5
Molecular Weight: 490-510
Color: 3-5R

HMW:
Sulfonate (%): 61-63
Water (%): 4-5
Molecular Weight: 510-550
Color: 4-5R

S:
Sulfonate (%): 61-63
Water (%): 4-5
Molecular Weight: 450-480
Color: 2.5-3.5R

STEPAN CO.: Agricultural Products:

Product:

TOXIMUL D:
  Sulfonate nonionic blend
  HLB: 10.5
  Appearance: Liquid
  Solubility: Xylene

TOXIMUL H:
  Sulfonate nonionic blend
  HLB: 13.5
  Appearance: Liquid
  Solubility: Xylene
  Applications: Emulsifier pair for insecticides.

TOXIMUL H-HF:
  Sulfonate nonionic blend
  HLB: 13.5
  Appearance: Liquid
  Solubility: Xylene
  Applications: High-flash version of TOXIMUL H.

TOXIMUL R:
  Sulfonate nonionic blend
  HLB: 10.5
  Appearance: Liquid
  Solubility: Xylene

TOXIMUL S:
  Sulfonate nonionic blend
  HLB: 13.0
  Appearance: Liquid
  Solubility: Xylene
  Emulsifier pair for herbicides.

TOXIMUL R-HF:
  Sulfonate nonionic blend
  HLB: 10.5
  Appearance: Liquid
  Solubility: Xylene
  High-flash version of TOXIMUL R.

TOXIMUL 351:
  Ether sulfate
  HLB: 15.0
  Appearance: Clear liquid
  Solubility: Xylene, water
  Emulsifier for pentachlorophenol

TOXIMUL 500(TOXIMUL 600):
  Sulfonate nonionic blend
  HLB: 10.5
  Appearance: Clear liquid
  Solubility: Xylene
  Emulsifier for insecticides

STEPAN CO.: Agricultural Products(Continued):

Product:

TOXIMUL MP:
  Sulfonate nonionic blend
  HLB: 11.0
  Appearance: Liquid
  Solubility: Xylene
  Applications: Emulsifier for insecticides

TOXIMUL MP-10:
  Sulfonate nonionic blend
  HLB: 12.0
  Appearance: Liquid
  Solubility: Xylene
  Applications: Emulsifier for insecticides. Use with TOXIMUL
D and H.

TOXIMUL MP-26:
  Sulfonate nonionic blend
  HLB: 12.5
  Appearance: Liquid
  Solubility: Xylene, water
  Applications: Emulsifier for high poundage organophosphates.

TOXIMUL 709:
  Sulfonate nonionic blend
  HLB: 10.5
  Appearance: Amber liquid
  Solubility: Xylene

TOXIMUL 710:
  Sulfonate nonionic blend
  HLB: 13.0
  Appearance: Amber liquid
  Solubility: Xylene
  Applications: Emulsifier pair for dinitro herbicides.

TOXIMUL 715:
  Sulfonate nonionic blend
  HLB: 10.5
  Appearance: Liquid
  Solubility: Xylene

TOXIMUL 716:
  Sulfonate nonionic blend
  HLB: 12.0
  Appearance: Liquid
  Solubility: Xylene
  Applications: Emulsifier pair for Dinitroaniline herbicides.

STEPAN CO.: Agricultural Products(Continued):

Product:

TOXIMUL 730:
  Nonionic blend
  HLB: 11.5
  Appearance: Liquid
  Solubility: Xylene
  Applications: Emulsifiers for crop oil concentrates based on vegetable oil.

TOXIMUL 804:
  Anionic blend
  HLB: 11.0
  Appearance: Amber liquid
  Solubility: Xylene
  Applications: Emulsifier for Propanil

TOXIMUL 811:
  Sulfonate nonionic blend
  HLB: 13.5
  Appearance: Amber liquid
  Solubility: Xylene, water
  Applications: Emulsifier for Betasan.

TOXIMUL 850M:
  Nonionic blend
  Appearance: Amber liquid
  Solubility: Xylene
  Applications: Emulsifier for paraffinic spray oils (17% use level).

TOXIMUL 852:
  Nonionic blend
  Appearance: Amber liquid
  Solubility: Xylene
  Applications: Emulsifier for vegetable oils (7-17% use level).

TOXIMUL 853:
  Nonionic blend
  Appearance: Liquid
  Solubility: Xylene
  Applications: Emulsifier for dormant spray oils (1% use level).

TOXIMUL 856:
  Nonionic blend
  Appearance: Clear liquid
  Solubility: Water
  Applications: Adjuvant-Nonionic spreader 80%

STEPAN CO.: Agricultural Products(Continued):

Product:

TOXIMUL 857:
  Nonionic blend
  Appearance: Clear yellow liquid
  Solubility: Water
  Applications: Adjuvant-Nonionic spreader-sticker 90%

TOXIMUL 858:
  Nonionic blend
  Appearance: Amber liquid
  Solubility: Xylene
  Applications: Adjuvant-Nonionic sticker

STEPFAC 8170:
  Anionic, acid form
  Appearance: Clear pale yellow liquid
  Solubility: Xylene, water
  Applications: Compatibility agent for liquid fertilizers 100%

STEPFAC 8171:
  Anionic, acid form
  Appearance: Clear pale yellow liquid
  Solubility: Xylene
  Applications: Compatibility agent.

STEPFAC 8172:
  Anionic, acid form
  Appearance: Clear pale yellow liquid
  Solubility: Xylene
  Applications: Compatibility agent.

STEPFAC 8173:
  Anionic, acid form
  Appearance: Clear pale yellow liquid
  Solubility: Xylene
  Applications: Compatibility agent.

TOXIMUL 8240:
  Castor Oil, POE-40
  HLB: 13.0
  Appearance: Amber liquid
  Solubility: Xylene, water
  Applications: Emulsifier component.

TOXIMUL 8241:
  Castor Oil, POE-30
  HLB: 12.0
  Appearance: Amber liquid
  Solubility: Xylene, water
  Applications: Emulsifier component

STEPAN CO.: Agricultural Products(Continued):

Product:

TOXIMUL 8301:
  Nonionic blend
  Appearance: Clear yellow liquid
  Solubility: Water
  Applications: Wetting agent--100% active.

TOXIMUL 8320:
  Butyl EO PO block copolymer
  HLB: 12.0
  Appearance: Light amber liquid
  Solubility: Xylene, water
  Applications: Emulsifier component flowable surfactant.

TOXIMUL 8321:
  Block copolymer
  HLB: 5.5
  Appearance: Amber liquid
  Solubility: Xylene
  Applications: Emulsifier component wetting agent.

TOXIMUL 8322:
  Block copolymer
  HLB: 14.0
  Appearance: Amber liquid to paste
  Solubility: Xylene
  Applications: Emulsifier component.

TOXIMUL 8323:
  Block copolymer
  HLB: 17.0
  Appearance: Amber liquid to paste
  Solubility: Xylene, water
  Applications: Emulsifier component flowable surfactant.

TOXIMUL TA-2:
  Tallow amine, POE
  HLB: 5.0
  Appearance: Amber liquid
  Solubility: Xylene
  Applications: Emulsifier component.

TOXIMUL TA-5:
  Tallow amine, POE
  HLB: 9.0
  Appearance: Amber liquid
  Solubility: Xylene
  Applications: Emulsifier component.

STEPAN CO.: Agricultural Products(Continued):

Product:

NIPOL 2782:
  Block polymer
  HLB: 14.0
  Appearance: Tan solid
  Solubility: Xylene, water
  Applications: Emulsifier component.
NIPOL 4472:
  Block polymer
  HLB: 12.5
  Appearance: Tan solid
  Solubility: Xylene, water
  Applications: Emulsifier component.
NIPOL 5595:
  Block polymer
  HLB: 15.0
  Appearance: Tan solid
  Solubility: Xylene, water
  Applications: Emulsifier component.
NIPOL 5690:
  Block polymer
  HLB: 15.0
  Appearance: Tan solid
  Solubility: Xylene, water
  Applications: Emulsifier component.
MICRO-STEP H-301:
  Sulfonate, nonionic blend
  Appearance: Amber liquid
  Solubility: Xylene, water
  Applications: Emulsifier for microemulsions.
MICRO-STEP H-302:
  Sulfonate, nonionic blend
  Appearance: Amber liquid
  Solubility: Xylene, water
  Applications: Emulsifier for microemulsions.
MICRO-STEP H-303:
  Nonionic blend
  Appearance: Amber liquid
  Solubility: Xylene, water
  Applications: Emulsifier for microemulsions
MICRO-STEP H-304:
  Sulfonate, nonionic blend
  Appearance: Amber liquid
  Solubility: Xylene, water
  Applications: Emulsifier for microemulsions.
MICRO-STEP H-305:
  Sulfonate, nonionic blend
  Appearance: Amber liquid
  Solubility: Xylene, water
  Applications: Emulsifier for microemulsions.

STEPAN CO.: Esters:

Emulsifiers, Opacifiers:

Product:

Glycerol Monostearate Pure (GMS):
   CTFA: Glyceryl Stearate
   HLB Value: 3.8
   Acid Value (Max.): 3.0
   Melting Point C: 56.5-58.5
   Emulsifier-opacifier and bodying agent. Used in creams, lotions, anti-perspirants, hair care products and sun screens.

Glycerol Distearate 386F (GDS):
   CTFA: Glyceryl Distearate
   HLB Value: 2.4
   Acid Value (Max.): 5.0
   Melting Point C: 55-60
   Alternative to Glycerol Monostearate Pure offering lower HLB value.

Glycerol Monostearate S.E. (GMS-S.E.):
   CTFA: Glyceryl Stearate SE
   Acid Value (Max.): 20
   Melting Point C: 56.5-59.5
   Anionic modified. Recommended for use in oil-in-water emulsions that are in the pH range of 5 to 9.

Glycerol Monostearate S.E. acid stable (GMS-S.E.A.S.):
   CTFA: Glyceryl Stearate (and) PEG-100 Stearate
   HLB Value: 11.2
   Acid Value (Max.): 3.0
   Melting Point C: 54-58
   Nonionic. Recommended for low pH (3 to 5). Used as emulsifier, self-emulsifying cream base, hair and skin conditioner. Good electrolyte stability.

Glycerol Monooleate (GMO):
   CTFA: Glyceryl Oleate
   HLB Value: 3.8
   Acid Value (Max.): 3.0
   Melting Point C: 20
   Effective water-in-oil emulsifier. Used in bath oil as emollient and spreading agent, in makeup as pigment dispersant and in vanishing and moisturizing cream to impart slip.

STEPAN CO.: Esters(Continued):

Product:

Emulsifiers, Opacifiers(Continued):

Glycerol Monolaurate (GML):
 CTFA: Glyceryl Laurate
 HLB: 4.9
 Acid Value (Max.): 5.0
 Melting Point C: 54
 Glyceryl Monolaurate functions as a primary emulsifier for water-in-oil emulsions. In addition to its emulsifier function, Glycerol Monolaurate can impart a lasting emollient feel to formulations.

Glycerol Dilaurate (GDL):
 CTFA: Glyceryl Dilaurate
 HLB Value: 4.0
 Acid Value (Max.): 5.0
 Melting Point C: 30
 Glycerol Dilaurate is a semi-solid recommended for use in free-flowing lotions where the glycerol laurate emolliency is desired.

Pearlescent Agents, Auxiliary Emulsifiers:

Ethylene Glycol Monostearate Pure (EGMS):
 CTFA: Glycol Stearate
 HLB Value: 2.9
 Acid Value (Max.): 2.0
 Melting Point C: 56-60
 Pearlescent agent in shampoos and liquid hand soaps. It also functions as a bodying agent and emulsion stabilizer in those systems.

Ethylene Glycol Distearate (EGDS):
 CTFA: Glycol Distearate
 HLB Value: 1.5
 Acid Value (Max.): 15.0
 Melting Point C: 60-63
 Pearlizer, emollient and emulsifier. Suggested for use when no additional viscosity response is desired such as high-solids formulations.

Ethylene Glycol Amido Stearate (EGAS):
 CTFA: Glycol Stearate (and) Stearamide AMP
 Acid Value (Max.): 5.0
 Melting Point C: 56.5-58.5
 Pearlescent and bodying agent in shampoos and liquid hand soaps. Also imparts a soft, smooth skin feel to formulations.

STEPAN CO.: Esters(Continued):

Pearlescent Agents, Auxiliary Emulsifiers(Continued):

Product:

Diethylene Glycol Monostearate (DGMS):
  CTFA: PEG-2 Stearate
  HLB Value: 4.3
  Acid Value (Max.): 5.0
  Melting Point C: 44.5-47.5
  Opacifier in shampoos and lotions. Imparts a luxurious emolliency and adds body to those formulations.

Propylene Glycol Monostearate Pure (PGMS):
  CTFA: Propylene Glycol Stearate
  HLB Value: 3.4
  Acid Value (Max.): 3.0
  Melting Point C: 33.5-38.5
  Melting point near that of body temperature and so is used in suppositories, lipsticks and sunscreens. Also functions as auxiliary emulsifier and opacifier.

Propylene Glycol Monolaurate (PGML):
  CTFA: Propylene Glycol Laurate
  HLB Value: 3.2
  Acid Value (Max.): 3.0
  Melting Point C: 33.5-38.5
  Light color and low odor liquid emollient and auxiliary emulsifier. Imparts a soft, velvety skin feel to cosmetic products.

Emulsifiers, Viscosity Builders:

PEG 200-6000 Mono and Dilaurates:
  CTFA: PEG-4 to PEG-150 Laurate and Dilaurate
  HLB Value: 5.9-19.3
  Acid Value (Max.): 5-10
  Melting Point C: 5-61

PEG 200-6000 Mono and Dioleates:
  CTFA: PEG-4 to PEG-150 Oleate and Dioleate
  HLB Value: 5.0-19.1
  Acid Value (Max.): 5-10
  Melting Point C: -15-59

PEG 200-6000 Mono and Distearates:
  CTFA: PEG-4 to PEG-150 Stearate and Distearate
  HLB Value: 4.8-19.1
  Acid Value (Max.): 5-10
  Melting Point C: 28-61

Nonionic emulsifiers covering wide HLB range. Non-toxic and non-irritants. Viscosity modifiers, emollients, opacifiers, spreading agents, wetting and dispersing agents.

STEPAN CO.: Personal Care Products: Cationics:

Stearyl Dimethyl Benzyl Ammonium Chloride:

Product:

AMMONYX 4:
    CTFA: Stearalkonium Chloride
    Active %: 17-19
    Physical Form: Paste

AMMONYX 4B:
    CTFA: Stearalkonium Chloride
    Active %: 16-18
    Physical Form: Paste

AMMONYX 4-IPA:
    CTFA: Stearalkonium Chloride
    Active %: 17-19
    Physical Form: Paste

AMMONYX 485:
    CTFA: Stearalkonium Chloride
    Active %: 85 min.
    Physical Form: Powder

AMMONYX 4002:
    CTFA: Stearalkonium Chloride
    Active %: 94 min.
    Physical Form: Powder

AMMONYX CA-Special:
    CTFA: Stearalkonium Chloride
    Active %: 20.0-22.5
    Physical Form: Paste
    These products possess pronounced conditioning, softening, and emolliency characteristics. Suggested applications include hair rinses, skin creams and lotions. Used as a cationic emulsifier.

Cetyl Trimethyl Ammonium Chloride:

AMMONYX CETAC:
    CTFA: Cetrimonium Chloride
    Active %: 24-26
    Physical Form: Liquid

AMMONYX CETAC-30:
    CTFA: Cetrimonium Chloride
    Active %: 29 Min.
    Physical Form: Liquid
    These products possess pronounced conditioning, softening, and emolliency characteristics. Suggested applications include hair rinses, skin creams and lotions. Used as a cationic emulsifier.

STEPAN CO.: Personal Care Products: Nonionics/Amphoterics:

Ethoxylated Alkanolamides:

Product:

AMIDOX C-2:
 CTFA: PEG-2 Cocamide
 Active %: 100
 Physical Form: Liquid

AMIDOX C-5:
 CTFA: PEG-5 Cocamide
 Active %: 100
 Physical Form: Liquid

AMIDOX L-2:
 CTFA: PEG-2 Lauramide
 Active %: 100
 Physical Form: Solid

AMIDOX L-5:
 CTFA: PEG-5 Lauramide
 Active %: 100
 Physical Form: Solid
 Mild, effective emulsifiers for fragrances and essential oils. Also, impart viscosity and foam enhancement.

Amine Oxides:

AMMONYX CO:
 CTFA: Cetamine Oxide
 Active %: 29-31
AMMONYX LO:
 CTFA: Lauramine Oxide
 Active %: 29-31
AMMONYX DMCD-40:
 CTFA: Lauramine Oxide
 Active %: 40-42
AMMONYX MO:
 CTFA: Myristamine Oxide
 Active %: 29-31
AMMONYX MCO:
 CTFA: Myristamine Oxide
 Active %: 29-31
AMMONYX SO:
 CTFA: Stearamine Oxide
 Active %: 24.5-31.5
AMMONYX CDO:
 CTFA: Cocamidopropylamine Oxide
 Active %: 29.5-31.5

Conditioner, emulsifier, viscosity modifier with wetting, foaming, and foam stabilization properties.

UNICHEMA CHEMICALS, INC.: ESTOL Esters:

Chemical Type: Esters produced from fatty acids, glycerine and other mono- and polyfunctional acids and alcohols
Major Industrial Applications: Plastics, lubricants, cosmetics, emulsifiers, filter-tip plasticisers, foundry binders, natural oil substitutes, synthetic lubricants

Oleates:

ESTOL 1400 methyl oleate:
    Acid value: 1.0
    Saponification value: 190-196
    Iodine value: 86-93

ESTOL 1402 methyl oleate:
    Acid value: 1.5
    Saponification value: 190-196
    Iodine value: 86-93

ESTOL 1405 n-butyl oleate:
    Acid value: 0.5
    Saponification value: 166-172
    Iodine value: 75-82

ESTOL 1406 isopropyl oleate:
    Acid value: 0.5
    Saponification value: 172-180
    Iodine value: 75-86

ESTOL 1407 glycerol monooleate:
    Acid value: 1.0
    Saponification value: 165-175
    Iodine value: 71-79

ESTOL 1414 isobutyl oleate:
    Acid value: 0.5
    Saponification value: 166-172
    Iodine value: 75-82

ESTOL 1415 isooctyl oleate:
    Acid value: 0.5
    Saponification value: 140-146
    Iodine value: 65-71

ESTOL 1427 trimethylol-propane trioleate:
    Acid value: 2.0
    Saponification value: 178-187
    Iodine value: 75-90

UNICHEMA CHEMICALS, INC.: ESTOL Esters(Continued):

Oleates:

ESTOL 1428 propylene glycol dioleate:
  Acid value: 2.0
  Saponification value: 184-192
  Iodine value: 82-93

ESTOL 1429 propylene glycol dioleate:
  Acid value: 2.0
  Saponification value: 184-192
  Iodine value: 74-83

ESTOL 1435 glycerol trioleate:
  Acid value: 2.0
  Saponification value: 190-196
  Iodine value: 85-93

ESTOL 1445 pentaerythritol tetraoleate:
  Acid value: 1.0
  Saponification value: 185-193
  Iodine value: 82-92

ESTOL 1446 neopentyl glycol dioleate:
  Acid value: 3.0
  Saponification value: 178-184

ESTOL 1447 polyethylene glycol 400 dioleate:
  Acid value: 5.0
  Saponification value: 120-130
  Iodine value: 53-65

Caprylates, caprates:

ESTOL 1526 propylene glycol dipcaprylate/-caprate:
  Acid Value: 0.1
  Saponification value: 315-335
  Iodine value: 1.0

ESTOL 1527 glycerol tricaprylate/-caprate:
  Acid value: 0.1
  Saponification value: 335-360
  Iodine value: 2.0

UNICHEMA CHEMICALS, INC.: ESTOL Esters(Continued):

Stearates:

ESTOL 1451 n-butyl stearate:
  Acid value: 0.2
  Saponification value: 170-180
  Iodine value: 1.0

ESTOL 1458 isooctyl stearate:
  Acid value: 0.2
  Saponification value: 144-152
  Iodine value: 1.0

ESTOL 1462 glycerol monostearate SE:
  Acid value: 3.0
  Saponification value: 156-170
  Iodine value: 3.0

ESTOL 1473 glycerol monostearate:
  Acid value: 3.0
  Saponificaton value: 168-184
  Iodine value: 3.0

ESTOL 1476 isobutyl stearate:
  Acid value: 0.2
  Saponification value: 168-176
  Iodine value: 1.0

ESTOL 1481 cetostearyl stearate:
  Acid value: 2.0
  Saponification value: 105-115
  Iodine value: 1.0

ESTOL 1482 methyl stearate:
  Acid value: 1.0
  Saponification value: 190-205
  Iodine value: 1.0

UNICHEMA CHEMICALS, INC.: ESTOL Esters(Continued):

Laurates, myristates, palmitates:

ESTOL 1502 methyl laurate:
  Acid value: 0.5
  Saponification value: 255-265
  Iodine value: 1.0

ESTOL 1503 methyl palmitate:
  Acid value: 1.0
  Saponification value: 205-209
  Iodine value: 0.1

ESTOL 1512 isopropyl myristate:
  Acid value: 0.1
  Saponification value: 206-211
  Iodine value: 0.5

ESTOL 1514 isopropyl myristate:
  Acid value: 0.5
  Saponification value: 206-211
  Iodine value: 1.0

ESTOL 1517 isopropyl palmitate:
  Acid value: 0.5
  Saponification value: 185-191
  Iodine value: 1.0

Others:

ESTOL 1501 methyl ester of tallow fatty acids:
  Acid value: 1.0
  Saponification value: 192-200
  Iodine value: 48-56

ESTOL 3075 diethyl phthalate(perfumery grade):
  Acid value: 0.07
  Saponification value: 501-509

UNICHEMA CHEMICALS, INC.: ESTOL Esters(Continued):

Acetates:

ESTOL 1574 ethylene glycol diacetate:
   Acid Value: 1.0
   Saponification value: 753-768

ESTOL 1579 glycerol triacetete(triacetin):
   Acid value: 0.05
   Saponification value: 765-774

ESTOL 1580 glycerol triacetate(triacetin):
   Acid value: 0.1
   Saponification value: 765-774

ESTOL 1581 glycerol triacetate(triacetin):
   Acid value: 0.5
   Saponification value: 755-774

ESTOL 1582 glycerol diacetate(diacetin):
   Acid value: 0.5
   Saponification value: 542-605

ESTOL 1583 glycerol diacetate(diacetin):
   Acid value: 0.5
   Saponification value: 525-550

ESTOL 1593 triethylene glycol diacetate(TEGDA):
   Acid value: 0.2
   Saponification value: 475-485

ESTOL 1594 triacetin TEGDA blend 1/1:
   Acid value: 0.2
   Saponification value: 620-630

ESTOL 1597 filter-tip adhesive:
   Acid value: 0.1
   Saponification value: 750-770

UNICHEMA CHEMICALS, INC.: PRICERINE Glycerines:

Chemical Type: Chemically pure and technical grades of glycerine

Major Industrial Applications: Pharmaceuticals, surface coating resins, nitration products, tobacco emulsifiers, cosmetics, esters

Product:

PRICERINE 9081 C.P., E.P.:
   Glycerine content % (min.): 86.5
   Water content %: <13.5
   Relative density: 1.228

PRICERINE 9083 C.P., E.P.:
   Glycerine content % (min.): 99.5
   Water content %: <0.5
   Relative density: 1.262

PRICERINE 9071 nitration grade:
   Glycerine content % (min.): 99.5
   Water content %: <0.5
   Relative density: 1.262

PRICERINE 9099 C.P.:
   Glycerine content % (min.): 99.7
   Water content %: 0.2 max.
   Relative density: 1.263

VAN DEN BERGH FOOD INGREDIENTS GROUP: DUR-EM Emulsifiers:

DUR-EM GMO:
Glycerol Monooleate

DUR-EM GMO glycerol monooleate is a surface active lipophilic agent made by the direct esterification of fatty acids and glycerol. It is used as an oil-soluble surfactant and lubricant in a number of industrial products.

Principal Uses:
The chemical properties of DUR-EM GMO glycerol monooleate permit it to be used as an emulsion stabilizer, solvent, wetting agent, and penetrant. It finds extensive application in cosmetic products, metal processing compounds, dry cleaning bases, paints and insecticide compositions.

Typical Data:
Alpha Mono Content: 42% Min.
Form: Plastic
Saponification Number: 160-170
Type: Nonionic
HLB: 2.8

DUR-EM 207E Emulsifier:

DUR-EM 207E is a finely powdered mono- and diglyceride specifically designed as a crumb softening agent in bread and other yeast-raised baked goods. It may also be used in other processes where a fine powder is required for dispersion. DUR-EM 207E requires no hydrating or melting for use. Its fine particle size allow it to disperse directly in both continuously and conventionally processed bread. Ease of handling, economy, and high functionality are recognized advantages of DUR-EM 207E.

Typical Data:
Alpha Monoglyceride Content: 50% (Min.)
A.O.M.: 200
Capillary Melting Point: 140-146F

Summary of Advantages:
Ease of Handling
Economy
High Functionality
Flexibility in Use

VAN DEN BERGH FOOD INGREDIENTS GROUP: DUR-EM Emulsifiers (Continued):

DUR-EM 300:
DUR-EM 300 emulsifier is a blend of mono- and diglycerides prepared by the direct esterification of fatty acids and glycerine with added propylene glycol to solubilize the mono- and diglycerides.

Principal Uses:
DUR-EM 300 emulsifier can be used in wide variety of applications. It is particularly applicable where physical characteristics of the finished product requires that the emulsifier system be liquid. DUR-EM 300 can be used as:
* Solubilizer or dispersing agent in flavor and color systems
* Antifog agent in polyester films.
* Emulsifier for hand creams and emollients. Provides lubricity to shaving cream lotions and cosmetic creams.

Typical Data:
Alpha Monoglyceride Content: 45% Min.
HLB: 2.8
Iodine value: 60-67
Propylene Glycol: 8% - 12%

DUR-EM 300K:
Emulsifier--Kosher

DUR-EM 300K is a blend of mono- and diglycerides prepared by the direct esterification of fatty acids and glycerine with added propylene glycol to solubilize the mono- and diglycerides.

Applications:
DUR-EM 300K can be used in a wide variety of applications. It is particularly applicable where physical characteristics of the finished product requires that the emulsifier system be liquid.
DUR-EM 300K can be used as:
* Solubilizer or dispersing agent in flavor and color systems.
* Antifoam agent in high-sugar such as juices and jellies, as well as high-protein systems.
* Antifog agent in polyester films.
* Wetting agent in spray-dried foods.
* Processing aid in the manufacture of yeast.

Typical Data:
Alpha Monoglyceride Content: 46% Min.
Iodine Value: 6 Min.
Propylene Glycol: 8% - 12%
Flavor: Bland

General Characteristics:
Total Monoglycerides (Alpha and Beta forms): Approximately 55%
HLB: 2.8

VAN DEN BERGH FOOD INGREDIENTS GROUP: DUR-EM Mono- and Diglycerides: Kosher:

DUR-EM emulsifiers are food grade mono- and diglycerides derived from vegetable fats and oils.

DUR-EM 114:
   Free Fatty Acids (% Oleic): 1.5 (Max.)
   Alpha Monoglycerides, %: 40 (Min.)
   Iodine Value: 65-75
   Capillary Melting Point, F: 110-120
   Antioxidants: Citric Acid, BHA
   HLB: 2.8
   Total Approx. Monoglycerides: 48

DUR-EM 117:
   Free Fatty Acids (% Oleic): 1.0 (Max.)
   Alpha Monoglycerides, %: 40 (Min.)
   Iodine Value: 5 (Max.)
   Capillary Melting Point, F: 140-150
   Antioxidants: Citric Acid
   HLB: 2.8
   Total Approx. Monoglycerides: 48

DUR-EM 204:
   Free Fatty Acids (% Oleic): 1.25 (Max.)
   Alpha Monoglycerides, %: 52 (Min.)
   Iodine Value: 65-75
   Capillary Melting Point, F: 114-121
   Antioxidants: Citric Acid, BHA
   HLB: 3.5
   Total Approx. Monoglycerides: 62

DUR-EM 207:
   Free Fatty Acids (% Oleic): 1.5 (Max.)
   Alpha Monoglycerides, %: 52 (Min.)
   Iodine Value: 5 (Max.)
   Capillary Melting Point, F: 140-146
   Antioxidants: Citric Acid
   HLB: 3.5
   Total Approx. Monoglycerides: 62

Uses:
   Applications for DUR-EM 114 and 204 include: Icings, Cakes, Margarines, Danish, Sweet Doughs, Whipped Toppings and other Vegetable Dairy Systems.

   DUR-EM 117 and 207 have found applications in: Margarines, Frozen Desserts, Whipped Toppings and other Vegetable Dairy Systems; Candies, Chewing Gums, Breads, Danish and Sweet Doughs.

VAN DEN BERGH FOOD INGREDIENTS GROUP: DURFAX Emulsifiers:

DURFAX EOM:
 Kosher

DURFAX EOM is a multifunctional surfactant composed of ethoxylated mono- and diglycerides (Polyglycerate 60) produced by reaction of ethylene oxide with a DUR-EM type monoglyceride. It is hydrophilic in nature, with an HLB of approximately 13.5, and exhibits a great degree of surface activity. DURFAX EOM is highly efficient and economical.

Applications:
 * Yeast raised products and shortenings
 * Cakes, cake mixes, and shortenings
 * Icings and creme fillers, icing mixes and shortenings
 * Hydrates

Typical Data:
 Acid Value: 0-2
 Hydroxyl Value: 65-80
 Saponification Value: 65-75
 Oxyethylene Content: 60.5%-65.0%
 Flavor: Bland

DURFAX 60:
 Polyoxyethylene (20) Sorbitan Monostearate Ester

DURFAX 60 polyoxyethylene sorbitan monostearate is a nonionic, water-dispersible surface active agent. Its chemical properties and compatibility with other materials make it an important ingredient in the manufacture of various types of specialty products.

Principal Uses:
 DURFAX 60 polyoxyethylene sorbitan monostearate is used in a number of cosmetic, toiletries and household products such as hand creams, shaving creams, deodorant creams and waterless hand cleaners. It is also used in industrial applications such as insecticide formulations, chemical dispersions and many others.

Typical Data:
 Type: Nonionic
 Saponification Number: 45-55
 Moisture: 3.0% Max.
 Form: Liquid
 Acid Value: 2.0 Max.
 Hydroxyl Value: 81-96
 HLB: 14.9

VAN DEN BERGH FOOD INGREDIENTS GROUP: Emulsifiers:

DURFAX 80 and DURFAX 80K:
Polyoxyethylene (20) Sorbitan Monooleate Ester

DURFAX 80 polyoxyethylene sorbitan monooleate is a nonionic, water dispersible surface active agent. It is used in hair shampoos; lotions; aerosol shave creams; as an anti-fog agent in plastics and aerosol furniture polishes.

Typical Data:
Acid Value: 3.0 Max.
Saponification Value: 45-55
Hydroxyl Value: 65-80
Iodine Value: 65.0
HLB: 15.9
Flavor: Slight, Characteristic
Clarity: Clear Yellow Liquid

DURLAC 100W:
Kosher

DURLAC 100W is composed of the lactic acid esters of mono- and diglycerides.

Applications:
DURLAC 100W is a very good alpha-tending emulsifier that is capable of bringing about an agglomeration of fat globules, which promotes the formation of good and stable aerated products.
DURLAC 100W is widely used within the food industry as an emulsifier and an aerating agent. Typical major application areas include:
* Cake shortenings to improve texture and volume in baked goods.
* Powdered desserts and whipped toppings to improve aeration and stability.
* Pudding mixes to improve consistency and stability.
* Cakes to impart optimum aeration effect and stability.
* Confectionery-type coatings to enhance gloss retention/ bloom retardation.
* Confectionery-type coatings to reduce snowflaking or mottling in molded coatings.

Typical Data:
Acid Value: 5.0 Max.
Capillary Melting Point: 115-130F (46-55C)
Alpha Monoglyceride: 10%
WICLA (Water Insoluble Combined Lactic Acid): 13% Min.

General Characteristics:
Saponification Value: 245-260
HLB: 2.4
Emulsifier Class/Type: Nonionic

VAN DEN BERGH FOOD INGREDIENTS GROUP: Emulsifiers(Continued):

DUR-LO:
  Mono- and Diglyeride Emulsifier for Fat-Reduced Foods

  DUR-LO is the DURKEE tradename for a vegatable oil mono- and diglyceride used to reduce the fat content in food formulations. Unlike other such emulsifiers, DUR-LO is low in alpha mono content and relatively high in diglyceride content. Functionally, it can replace all or part of the shortening content in cake mixes, cookies, icings, and numerous vegetable dairy products. It also can reduce calories in these products.

  Advantages:
    * Replaces or reduces fat content
    * Reduces calories
    * Matches or reduces lipid costs
    * Allows removal of "fat" term from product label

  Typical Data:
    Capillary Melting Point: 113-120F (45-49C)
    Iodine Value: 66-70
    Alpha Mono: 17% - 22%
    Color/Form: Ivory/Plastic
    Flavor/Odor: Good

EC-25:
  Emulsifier system for prepared cake mixes, cakes, cookies and shortenings

  EC-25 is an emulsifier system designed for use in prepared cake mixes, cakes, cookies and shortenings. This emulsifier system improves moisture retention, improves air incorporation, provides fine grain, good tenderness, and improved volume in finished cakes. The EC-25 system permits ease of handling in incorporating into cake mixes. This one ingredient provides the advantages of several emulsifiers when added to cake mixes.

  It can also be used in shortenings when manufacturing soft-type cookies and very lean cakes

  Typical Data:
    Alpha Monoglyceride Content: 20%-25%
    Propylene Glycol Monoester (PGME): 34%-38%
    Capillary Melting Point: 90-100F
    Color: Straw

VAN DEN BERGH FOOD INGREDIENTS GROUP: Emulsifiers(Continued):

DURFAX 65:
   Color & Form: Tan Solid
   Label Ingredient Statement: Polysorbate 65
   Typical Data:
      Sap. Value: 88-98
      Hyd. Value: 44-60
   FDA Ref. # (21CFR): 172.838

DURTAN 60:
   Color & Form: Cream Flake
   Label Ingredient Statement: Sorbitan Monostearate
   Typical Data:
      Sap. Value: 147-157
      Hyd. Value: 235-260
   FDA Ref. # (21CFR): 172.842

DURKEE 400MO: **
   Color & Form: Amber Semi-solid
   Label Ingredient Statement: Polyethylene Glycol (400) Mono- and Dioleate
   Typical Data:
      Sap. Value: 80-88
      Hyd. Value: 80-93
   FDA Ref. # (21CFR): 573.820**
                      573.800**
   Applications: In calf milk replacers, processing aid in animal feeds.

** Permitted in Feed and Drinking Water of Animals

DURPRO 107:
   Color & Form: Light Straw Flakes
   Label Ingredient Statement: Propylene Glycol Mono- and Diesters of Fats and Fatty Acids with Citric Acid to help protect flavor.
   Typical Data:
      PGME: 50% (Min.)
      Mono: 10% (Min.)
      CMP: 115-125F
   FDA Ref. # (21CFR): 172.856
   Applications: Cakes: Mixes, Used with all-purpose shortenings for baker's cakes or with emulsified shortenings for very lean cakes.

VAN DEN BERGH FOOD INGREDIENTS GROUP: Emulsifiers(Continued):

Emulsifiers--Frozen Desserts:

ICE #2:
   Color & Form: Light Ivory Bead
   Label Ingredient Statement: Mono- and Diglycerides and 20% Polysorbate 80
   Typical Data:
      Mono: 32%-38%
      CMP: 138-144F
  FDA Ref. # (21CFR): 182.4505 and 172.840
   Applications: Used as an emulsifier in frozen desserts

ICE #12:
   Color & Form: Light Ivory Bead
   Label Ingredient Statement: Mono- and Diglycerides and 20% Polysorbate 65
   Typical Data:
      Mono: 32%-36%
      CMP: 138-144F
  FDA Ref. # (21CFR): 182.4505 and 172.838
   Applications: Used as an emulsifier in frozen desserts

ICE #81:
   Color & Form: Bead
   Label Ingredient Statement: Mono- and Diglycerides and Polyglycerol Esters of Fatty Acids
   Typical Data:
      Mono: 34%-40%
      CMP: 135-144F
  FDA Ref. # (21CFR): 182.4505, 172.854, 135.110(C) and 135.140(C)
   Applications: Used as an emulsifier in frozen desserts

VAN DEN BERGH FOOD INGREDIENTS GROUP: SANTONE Emulsifiers:

SANTONE 3-1-S XTR:
Triglycerol Monostearate: Kosher
* Lighter Color
* Improved Flavor
* Extra Functionality

SANTONE 3-1-S XTR is the polyglycerol ester, triglycerol monostearate, designed for use as a whipping agent, as an aerator in non-aqueous lipid systems and providing emulsion stability.

Applications:
SANTONE 3-1-S XTR can be used with various ingredients to formulate an unlimited number of low-density, shelf-stable products: candy centers, fillings, toppings, spreads, and meltable ingredients in semi-prepared foods. SANTONE 3-1-S XTR can be used to prepare water in oil and oil in water emulsions.

Typical Data:
Saponification Number: 130-145
Capillary Melting Point: 125-135F
Mettler Dropping Point: 52-57C
Hydroxyl Number: 300-350
HLB: 6.9

SANTONE 8-1-O:
Octaglycol Monooleate: Kosher
SANTONE 8-1-O is the polyglycerol ester, octaglycerol monooleate, for use as a replacement for polysorbates, as a viscosity reducer in high protein systems, as an emulsion stabilizer, and as a beverage clouding agent.

Applications:
* Polysorbate Replacer
* Whipped Toppings
* Color and Flavor Dispersers
* Protein Systems
* Emulsion Stabilizers
* Clouding Agent
* Ice Cream Toppings
* Chocolate and Compound Confections
* Wetting Aid
* Food Analogs

Typical Data:
Saponification Number: 75-85
Hydroxyl Number: 495-550
Iodine Value: 30-40
HLB: 13.0
Viscosity at 50C = 16000 cps

VAN DEN BERGH FOOD INGREDIENTS GROUP: TALLY 100 Dough Conditioners and Softeners:

For Yeast-Raised Baked Goods

TALLY is the tradename for several dough conditioners and softeners developed to meet the diverse needs of bakers of yeast-raised products. Products include TALLY 100K PLUS, and TALLY 100K PLASTIC. TALLY products are composed of mono- and diglycerides and ethoxylated mono- and diglycerides. They are specifically designed as economical dough conditioners and softeners in yeast-raised baked goods. They provide tolerance to mixing variation, excellent resistance to shock after proofing, improved volume, and superior softening properties.

Applications:
  Conventional Process Bread
  Continuous Process Bread
  Sweet Doughs

Features:
  TALLY products are available in beaded or plastic form and in varying ratios of mono- and diglycerides and ethoxylated mono- and diglycerides. Each of these products possess the above mentioned advantages.

TALLY 100K PLUS:
  A free flowing beaded product developed to provide the maximum dough conditioning effects while simultaneously providing anti-staling properties in yeast baked goods. Contains high levels of alpha monoglyceride integrated with 32% ethoxylated mono- and diglycerides.

TALLY 100K PLASTIC:
  A cream-colored, plasticized product noted for high dough conditioning properties and anti-staling features. TALLY 100K PLASTIC can be used in danish pastry and sweet doughs for improved machinability. Also ideal for those types of bread manufacture where a plastic product is desired. Contains 40% EO-mono.

Advantages:
  * Excellent Functionality
  * Speeds Processing
  * Tally also provides drier, more machinable doughs
  * Improves Quality
  * Outstanding Cost Savings

Legal:
  The use of TALLY 100 dough conditioner and softener is covered under U.S. Patent No. 3,433,645.

TALLY 100K PLUS:
  Alpha Monoglyceride Content: 33%-37%
  Capillary Melting Point: 130-137F
  Free Fatty Acid: 2.0 Max.

TALLY 100K PLASTIC:
  Alpha Monoglyceride Content: 30%-34%
  Capillary Melting Point: 118-125F
  Free Fatty Acid: 2.0 Max.

WESTVACO CORP.: Asphalt Chemicals:

Product Name:

INDULIN C:
Highly effective anionic stabilizer with excellent dispersant qualities for "oil in water" type asphalt emulsions, where rosin and fatty acid soaps are used as the primary emulsifier.

POLYFON H:
The dispersant-stabizer can be used in place of INDULIN C in asphalt emulsion formulations where a higher degree of dispersion is required.

INDULIN SA-L:
Developed as a primary emulsifier for use in slow-set anionic asphalt emulsions. Also can be used as a stabilizer in medium or rapid-setting grades of asphalt emulsions, and as a retarder in quick-setting anionic emulsions.

INDULIN W-1:
Acid-soluble lignin-amine primary emulsifier used in slow-set, cationic asphalt emulsions as designated in ASTM D-397-71.

INDULIN W-3:
Blended lignin-amine asphalt emulsifiers for use in cationic emulsions requiring improved storage and stability characteristics.

INDULIN AS-1:
Conventional amine-based, anti-stripping additive for use in a wide range of highway paving applications.

INDULIN AS-Special:
Amine-based, super-concentrated anti-stripping additive that makes an excellent asphalt adhesion agent.

INDULIN RK Series:
Amine-based emulsifiers specifically designed for cationic rapid-setting emulsions.

INDULIN MQK:
Cationic mixing-grade asphalt emulsifier based on Westvaco's unique capabilities in the area of fatty acid chemistry. Forms a stable asphalt emulsion which minimizes storage and shipping problems and is effective with most types of emulsion-based asphalts. Also for slurry seal applications.

WESTVACO CHEMICALS INC.: Asphalt Chemicals(Continued):

Product Name:

INDULIN MQK-1M:
Like INDULIN MQK, this product is specifically designed for hard-to-emulsify asphalts whose primary usage is conventional slurry seal.

INDULIN XD-70:
Versatile liquid coemulsifier that can be used for either anionic or cationic slow-setting asphalt emulsions.

INDULIN 814:
Organic polyamine-based adhesion promoter for use in asphalt emulsions.

INDULIN 201/202:
Poly basic fatty acids which function as emulsifiers in high float, rapid-setting, emulsions (HFRS).

INDULIN 206:
Additive which functions as a high float promoter in HFRS emulsions.

INDULIN AQS/AQS-1M:
Unique, anionic quick-set emulsifiers designed for slurry seal applications.

WESTVACO CHEMICALS INC.: Rubber Chemicals:

Stabilized Rosin Acids

Product Name:

WESTVACO Resin 90:
For rubber producers who prepare their own soaps, Resin 90 is an excellent base for polymerization reaction emulsifiers. Its high purity ensures stability and consistency in the polymer and its fluidity increases handling safety and energy savings by lowering minimum handling temperatures. Its fluidity, compatibility, and heat stable properties also make it suitable for use as a shortstop in various solution polymers. Resin 90 can be used as a plasticizer and tackifier in any application requiring a low softening point, stabilized rosin.

WESTVACO Resin 790:
Sodium soap of Resin 90.

WESTVACO Resin 95:
Designed for use as an emulsifier in SBR polymerization. Its high purity ensures stability and consistency in the polymer and its fluidity increases handling safety and energy savings by lowering minimum handling temperatures. Its fluidity, compatibility, and heat stable properties also make it suitable for use as a shortstop in various solution polymers, where it provides tack and stability to the finished polymer. This versatility enables the synthetic rubber producer to use one basic tackifier in both emulsion polymerization polymers and solution polymers.

WESTVACO Resin 795:
Sodium soap of Resin 95.

WESTVACO Resin 895:
Potassium soap of Resin 95.

Stabilized Mixed Acids:

WESTVACO M-30:
       M-40:
       M-70:

Blends of catalytically stabilized tall oil rosin and fatty acids. M-30 contains approximately 30% stabilized rosin; M-40 contains 40%; M-70 contains 70%. Used primarily as an emulsifier in the polymerization of styrene and butadiene for synthetic tire rubber, and other applications requiring a blend of catalytically stabilized fatty acids and rosin acids.

300 Emulsifying Agents

WITCO CORP.: Surfactants for Emulsion Polymerization: General Emulsifier Recommendations:

Surfactants:

WITCOLATE SL-1:
  Monomer Systems: Acrylics, acrylonitrile, carboxylated SBR, chloroprene, styrene, vinyl chloride, vinyl acetate
  General Performance Characteristics: Gives small latex particle size and moderately stable monomer emulsions with moderately-low coagulum. HLB about 40.
  Chemical Description: Sodium Lauryl Sulfate

WITCONATE AOS:
  Monomer Systems: Acrylics, acrylonitrile, chloroprene, styrene, vinyl acetate, vinyl acrylics, SBR, carboxylated SBR, vinyl chloride, vinyl chloride copolymers.
  General Performance Characteristics: Gives small latex particle size and stable monomer emulsiions and polymer latices with low coagulum. Produces best results of surfactants tested in model styrene and vinyl acrylic systems. Stable at high temperatures and over a wide pH range. HLB about 11.8.
  Chemical Description: Sodium C14-16 Olefin Sulfonate

WITCONATE 1223H:
  Monomer Systems: ABS, SAN, SBR, styrene, vinyl acrylics, vinyl chloride
  General Performance Characteristics: Stable at high temperatures and over a wide pH range. Has low interfacial tension with many monomers and adds good wetting properties to latices. Also used in vinyl acrylics with WITCONOL NP-330.
  Chemical Description: Sodium Branched Dodecylbenzene Sulfonate

WITCOLATE D51-51:
WITCOLATE D51-53:
WITCOLATE D51-60:
WITCOLATE D51-53HA:
  Monomer Systems: Acrylics, vinyl acetate, vinyl acrylics, styrene, SAN, styrene acrylics, vinyl chloride.
  General Performance Characteristics: Range of ethylene oxide adducts provides good long-term stability without the use of nonionics. Provide moderately large particle-size latices with broad distribution and improved water resistance of film. Best performance exhibited at a pH between 4 and 10. High-ring sulfonation of WITCOLATE D51-53HA improves latex and freeze-thaw stability.
  Chemical Description: Nonyl Phenol Ethoxy(4)Sulfate
                       Nonyl Phenol Ethoxy(10)Sulfate
                       Nonyl Phenol Ethoxy(30)Sulfate
                       Nonyl Phenol Ethoxy(10)Sulfate

WITCO CORP.: Surfactants for Emulsion Polymerization: General Emulsifier Recommendations(Continued):

Surfactants:

EMCOL 4300:
Monomer Systems: Acrylics, vinyl acetate, vinyl acrylics
General Performance Characteristics: Produces latices with moderately small particle size and good long-term stability. Has excellent "preemulsion" stability. Dual charge provides good buffering properties.
Chemical Description: Disodium Alcohol Ether Sulfosuccinate

EMCOL K-8300:
Monomer Systems: Acrylics, acrylic copolymers, styrene acrylics, vinyl acrylics
General Performance Characteristics: Produces small particle-size latices. Has good buffering properties, good monomer emulsion and latex long-term stability. As a cosurfactant improves compatibility with filler and pigment. Works well with WITCONATE AOS, WITCONATE D51-53 and/or WITCONOL NP-100.
Chemical Description: Disodium Oleamido-MIPA Sulfosuccinate

EMCOL DOSS:
Monomer Systems: Acrylonitrile
General Performance Characteristics: Good surfactant. Improves surface wetting. Has very low surface and interfacial tension.
Chemical Description: Sodium Dioctyl Sulfosuccinate

EMCOL 4910:
Monomer Systems: Acrylics, vinyl acrylics
Good Performance Characteristics: Unique low-foaming asymmetrical cosurfactant. Aids in formation of small particle-size latices. Terminal hydroxyl may enhance crosslinking.
Chemical Description: Sodium Lauryl/Propoxy Sulfosuccinate

EMPHOS CS-141:
EMPHOS CS-136:
Monomer Systems: Acrylics, vinyl acetate, vinyl acrylics
General Performance Characteristics: May be used individually or together to improve latex long-term stability, corrosion resistance and adhesion to metals and provide antistatic properties. Stable to hydrolysis at high temperatures and wide pH range.
Chemical Description: Nonyl Phenol Ethoxy(10)Phosphate Ester
              Nonyl Phenol Ethoxy(6)Phosphate Ester

WITCONOL NP-100:
Monomer Systems: Vinyl Acrylics
General Performance Characteristics: When used with WITCONATE AOS improves stability and helps control particle size.
Chemical Description: Nonyl Phenol Ethoxy(10)

WITCONOL NP-330:
Monomer Systems: ABS, SAN, SBR, Styrene, vinyl acrylics, vinyl chloride
General Performance Characteristics: Unique long-chain anhydrous liquid used as a cosurfactant with an anionic to improve latex long-term and freeze-thaw stability. Acts as a plasticizer.
Chemical Description: Nonyl Phenol Ethoxy/Propoxy(30)

SUPPLIERS' ADDRESSES

Akzo Chemicals, Inc.
300 South Riverside Plaza
Chicago, IL 60606
(312)-906-7500/(800)-828-7929

Alcolac
1099 Winterson Road
Linthicum, MD 21090
(301)-859-4900

Alkaril Chemicals Ltd.
3265 Wolfedale Road
Mississauga, Ontario,
Canada L5C IV8
(416)-270-5534/(800)-387-5940

Amerchol Corp.
P. O. Box 4051
Edison, NJ 08818-4051
(201)-287-1600

American Cyanamid Co.
One Cyanamid Plaza
Wayne, NJ 07470
(201)-831-2000/(800)-443-0443

Angus Chemical Co.
2211 Sanders Road
Northbrook, IL 60062
(312)-498-6700/(800)-323-6209

BASF Corp.
Chemicals Division
100 Cherry Hill Rd.
Parsippany, NJ 07054
(201)-316-3000/(800)-526-1072

Capital City Products Co.
P. O. Box 569
Columbus, OH 43216-0569
(614)-299-3131/(800)-848-1340

CasChem, Inc.
40 Avenue A
Bayonne, NJ 07002
(201)-858-7900/(800)-277-2436

Central Soya
Chemurgy Division
1300 Fort Wayne National Bank
  Bldg.
Fort Wayne, IN 46801-1400
(800)-552-0980/(800)-348-0960

W. A. Cleary Products, Inc.
P. O. Box 10
1049 Somerset St.
Somerset, NJ 08873
(201)-247-8000/(800)-524-1662

Climax Performance Materials
7666 W. 63rd St.
Summit, IL 60501
(312)-458-8450

Cyclo Chemicals Corp.
7500 N.W. 66th St.
Miami, FL 33166
(305)-592-6700

DeSoto, Inc.
2001 North Grove
P. O. Box 2199
Fort Worth, TX 76113
(817)-625-2111

DuPont Co.
Chemicals and Pigments Dept.
Wilmington, DE 19898
(302)-774-2099/(800)-441-9442

Eastman Chemical Products, Inc.
P. O. Box 431
Kingsport, TN 37662
(800)-327-8626

FMC Corp.
Marine Colloids Division
2000 Market St.
Philadelphia, PA 19103
(215)-299-6178/(800)-346-5101

Goldschmidt Chemical Corp.
914 E. Randolph Road
P. O. Box 1299
Hopewell, VA 23860
(804)-541-8658/(800)-446-1809

Grindsted Products, Inc.
P. O. Box 26
201 Industrial Parkway
Industrial Airport, KS 66031
(913)-761-8100

Gumix International, Inc.
2160 North Central Road
Fort Lee, NJ 07024-7552
(201)-947-6300/(800)-2-GUMIX-2

Harcros Chemicals Inc.
5200 Speaker Rd.
Kansas City, KS 66106-1095
(913)-321-3131

Henkel Corp.
300 Brookside Ave.
Ambler, PA 19002
(215)-628-1476/(800)-531-0815

Hercules Inc.
Hercules Plaza
Wilmington, DE 19894
(800)-247-4372

Humko Chemical Division
Witco
P. O. Box 125
Memphis, TN 38101
(901)-320-5800/(800)-624-2964

ICI Specialty Chemicals
Wilmington, DE 19897
(302)-575-3000

Jordan Chemical Co.
1830 Columbia Ave.
Folcroft, PA 19032
(215)-583-7000/(800)-523-7425

Lanaetex Products
151-157 Third Ave.
Elizabeth, NJ 07206
(201)-351-9700

Lipo Chemicals Inc.
207 19th Ave.
Paterson, NJ 07504
(201)-345-8600

Lonza Inc.
22-10 Route 208
Fair Lawn, NJ 07410
(201)-794-2400/(800)-526-7850

Mayco Oil & Chemical Co.
775 Louis Drive
Warminster, PA 18974
(215)-672-6600/(800)-523-3903

Mazer Chemicals
3938 Porett Drive
Gurnee, IL 60031
(312)-244-3410/(800)-323-0856

Milliken Chemical
Iron Ore Road
P. O. Box 1927
Spartanburg, SC 29304
(803)-573-2259

Mobay Corp.
Mobay Road
Pittsburgh, PA 15205-9741
(412)-777-2000/(800)-662-2927

NL Chemicals, Inc.
Spencer Kellogg Products
P. O. Box 210
Buffalo, NY 14225
(716)-626-2090

Patco Products
3947 Broadway
Kansas City, MO 64111
(816)-561-9050/(800)-821-2250

Quantum Chemical Corp.
Emery Division
11501 Northlake Drive
Cincinnati, OH 45249
(513)-530-7300/(800)-543-7370

Sandoz Chemicals Corp.
4000 Monroe Rd.
Charlotte, NC 28205
(704)-331-7000/(800)-631-8077

ScanRoad Inc.
P. O. Box 7677
Waco, TX 76714-7677
(817)-772-7677

Shell Chemical Co.
1 Shell Plaza
Houston, TX 77002
(713)-241-6161

Werner G. Smith, Inc.
1730 Train Ave.
Cleveland, OH 44113
(216)-861-3676

Sonneborn Division
Witco Corp.
520 Madison Ave.
New York, NY 10022-4236
(212)-605-3908/(800)-634-4010

Stepan Co.
22 Frontage Road
Northfield, IL 60093
(312)-446-7500

Unichema Chemicals, Inc.
4650 South Racine Ave.
Chicago, Il 60609
(312)-376-9000/(800)-833-2864

Van den Bergh Food Ingredients
925 Euclid Ave.
Suite 800
Cleveland, OH 44115
(216)-344-8500

Westvaco
Chemical Division
Box 70848
Charleston Heights, SC 29415-0848
(803)-740-2300

Witco Corp.
Organics Division
Box 45296
Houston, TX 77245
(713)-433-7281/(800)-231-1542

CHEMICAL NAME INDEX

Chemical Name	Trade Name	Supplier
Acetylated lanolin alcohol*	LANAETEX	Lanaetex
Acetylated monoglycerides	CETODAN	Grindsted
Acetylated monoglycerides	MYVACET	Eastman
Acrylic graft copolymer solution	HYPERMER	ICI Specialty Chemicals
Agar agar		Gumix
Alcohol alkoxylates	ALKASURF	Alkaril
Alcohol ethoxylate	ALKASURF	Alkaril
Alcohol phosphates	ZELEC	DuPont
Aliphatic phosphate esters	ALKAPHOS	Alkaril
Alkylamine-guanidine polyoxyethanol	AEROSOL	American Cyanamid
Alkylaryl ethoxylate carboxylic acid*	SANDOPAN	Sandoz
Aluminum stearate	DEHYMULS	Henkel
Amine-based emulsifiers	INDULIN	Westvaco
Amine salt of lauryl sulfate	DUPONOL	DuPont
Amine sulfonate-nonionic ethoxylate blends	ALKAMULS	Alkaril
Amino ethyl imidazoline-tall oil	ALKAZINE	Alkaril
2-Amino-2-ethyl-1,3-propanediol	AEPD	Angus
2-Amino-2-methyl-1-propanol	AMP	Angus
Ammonium salt of a sulfated alkylphenoxypoly(ethyleneoxy) ethanol	SIPEX	Alcolac
Ammonium salt of sulfated nonylphenoxypoly(ethyleneoxy) ethanol	AEROSOL	American Cyanamid
Anionic	CASUL	Harcros Chemicals
Anionic	TRYLON	Quantum
Anionic, acid form	STEPFAC	Stepan
Anionic blends	T-MULZ	Harcros Chemicals
Anionic fluorosurfactants	ZONYL	DuPont
Anionic-nonionic blends	ALKAMULS	Alkaril
Anionic/nonionic blends	T-MULZ	Harcros Chemicals
Anionic quick-set emulsifiers	INDULIN	Westvaco
Anionic stabilizer	INDULIN	Westvaco
Anionic surfactant	ABEX	Alcolac
Aromatic phosphate esters	ALKAPHOS	Alkaril
Aryl polyoxyether	SYN FAC	Milliken
Barium petroleum sulfonate	BARIUM PETRONATE	Sonneborn
Beef tallow	MYVEROL	Eastman
Beeswax	DEHYMULS	Henkel
Block copolymer	TOXIMUL	Stepan
Block copolymer, ABA type	HYPERMER	ICI Specialty Chemicals
Block polymer	NIPOL	Stepan
n-Butyl oleate	ESTOL	Unichema
n-Butyl stearate	ESTOL	Unichema
Calcium DBS	ALKASURF	Alkaril
Calcium petroleum sulfonate	PETRONATE	Sonneborn
Calcium stearoyl-2-lactylate	VERV	Patco Products
Calcium sulfonate-nonionic ethoxylate blends	ALKAMULS	Alkaril
Capric acid*	NEO-FAT	Akzo Chemical
Capric acid	CAPMUL	Capital City
Caprylic acid*	NEO-FAT	Akzo Chemical
Caprylic acid	CAPMUL	Capital City
Caprylic capric triglyceride*	MAZOL	Mazer
Carboxylated, C4 paraffinic ethoxylate	SANDOPAN	Sandoz
Carob gum		Gumix

*Cosmetic, Toiletry and Fragrance Association (CTFA) name.

Chemical Name	Trade Name	Supplier
Carrageenan	GELCARIN, LACTARIN, SEAGEL, SEAKEM, SEASPEN, VISCARIN	FMC
Castor oil	TOXIMUL	Stepan
Castor oil ethoxylate	DESONIC	DeSoto
Cationic asphalt emulsifier	CATIMULS	ScanRoad
Cationic asphalt emulsifier	INDULIN	Westvaco
Cationic fluorosurfactant	ZONYL	DuPont
Cetamine oxide	AMMONYX	Stepan
Ceteareth	CREMOPHOR	BASF
Ceteareth	EUMULGIN	Henkel
Ceteareth*	LIPOCOL	Lipo
Ceteareth*	MACOL	Mazer
Cetearyl alcohol and ceteareth	CYCLOCHEM	Alcolac
Cetearyl alcohol and ceteareth	PROMULGEN	Amerchol
Cetearyl alcohol and ceteareth*	LIPOWAX	Lipo
Cetearyl alcohol and ceteth and glycol stearate*	CYCLOCHEM	Alcolac
Cetearyl alcohol and polysorbate*	LIPOWAX	Lipo
Ceteth*	LANYCOL	Lanaetex
Ceteth*	LIPOCOL	Lipo
Ceteth*	MACOL	Mazer
Cetostearyl stearate	ESTOL	Unichema
Cetrimonium chloride*	AMMONYX CETAC	Stepan
Cetyl acetate	SOLULAN	Amerchol
Cetyl esters*	CYCLOCHEM	Cyclo
Cetyl lactate	CYCLOCHEM	Cyclo
Cetyl palmitate*	CYCLOCHEM	Cyclo
Choleth*	ETHOXYCHOL	Lanaetex
Choleth and ceteth	SOLULAN	Amerchol
Cocamide*	ARMID	Akzo Chemical
Cocamide DEA*	MAZAMIDE	Mazer
Cocamidopropylamine oxide*	MAZOX	Mazer
Cocamidopropylamine oxide	AMMONYX	Stepan
Cocamidopropyl betaine*	MAFO	Mazer
Cocamidopropyl dimethylamine*	MAZEEN	Mazer
Cocamidopropyl hydroxysultaine*	MAFO	Mazer
Coco amido propyl dimethyl amine oxide	ALKAMOX	Alkaril
Coco betaine*	MAFO	Mazer
Coconut acid*	NEO-FAT	Akzo Chemical
Coconut acid	INDUSTRENE	Humko
Coconut diethanolamide	EMID	Quantum
Coconut monoethanolamide	EMID	Quantum
Coconut super diethanolamide	EMID	Quantum
Cottonseed oil	MYVEROL, MONOSET	Eastman
Diactyl tartaric acid esters of monoglycerides	PANODAN	Grindsted
Dibehenyl dimonium chloride	KEMAMINE	Humko
Diethanolamine salt of lauryl sulfate	DUPONOL	DuPont
Diethylene glycol monostearate	TEGIN	Goldschmidt
Diethyl phthalate	ESTOL	Unichema
2-Dimethylamino-2-methyl-1-propanol	DMAMP	Angus
Disodium alcohol ether sulfo-succinate	EMCOL	Witco Chemical
Disodium ethoxylated alcohol half ester of sulfosuccinic acid	AEROSOL	American Cyanamid
Disodium ethoxylated nonyl phenol half ester of sulfosuccinic acid	AEROSOL	American Cyanamid

Chemical Name	Trade Name	Supplier
Disodium isodecylsulfosuccinate	AEROSOL	American Cyanamid
Disodium mono- and didodecyl-diphenyloxide	AEROSOL	American Cyanamid
Disodium N-octadecyl sulfosuccinamate	AEROSOL	American Cyanamid
Disodium oleamide-MIPA sulfosuccinate	EMCOL	Witco Chemical
Dispersant/emulsifiers	SYN FAC	Milliken
Dispersant-stabilizer	POLYFON	Westvaco
Dispersing agent from edible tallow	MYVATEM	Eastman
Dispersing agents from oils	MYVATEM	Eastman
Dodecyl phenol ethoxylate	DESONIC	DeSoto
EGMS	CYCLOSHEEN	Cyclo
Ester	DESOTAN	DeSoto
Ether sulfate	TOXIMUL	Stepan
Ethoxylated amine	DESOMEEN	DeSoto
Ethoxylated amine	SYNFAC	Milliken
Ethoxylated castor oil	ALKASURF	Alkaril
Ethoxylated ester nonionic blends	ALKAMULS	Alkaril
Ethoxylated fatty acid	TRYDET	Quantum
Ethoxylated lauryl alcohol	TRYCOL	Quantum
Ethoxylated mixed rosin and fatty acids	TRYDET	Quantum
Ethoxylated tall oil	TRYDET	Quantum
Ethoxylate, EO	NEODOL	Shell Chemical
Ethylene glycol diacetate	ESTOL	Unichema
Ethylene glycol monostearate	TEGIN	Goldschmidt
Ethylene oxide derivative of hydrogenated castor oil	TAGAT	Goldschmidt
Fatty acid ester, lauric	ALKAMULS	Alkaril
Fatty acid ester, oleic	ALKAMULS	Alkaril
Fatty acid ester, stearic	ALKAMULS	Alkaril
Fatty acid ester, tallow	ALKAMULS	Alkaril
Fatty acid ethoxylate-lauric	ALKASURF	Alkaril
Fatty acid ethoxylate-oleic	ALKASURF	Alkaril
Fatty acid ethoxylate-palmitic	ALKASURF	Alkaril
Fatty acid ethoxylate-pelargonic	ALKASURF	Alkaril
Fatty acid ethoxylate-stearic	ALKASURF	Alkaril
Fatty acids and alcohols	CYCLOCHEM	Cyclo
Fatty diethanol amide	EMULAMID	Mayco
Glycerine	PRICERINE	Unichema
Glycerol diacetate (diacetin)	ESTOL	Unichema
Glycerol dioleate	ALKAMULS	Alkaril
Glycerol dioleate	EMEREST	Quantum
Glycerol esters	ATMOS	Humko
Glycerol mono distearates mixed with cationic co-emulsifiers	TEGINACID	Goldschmidt
Glycerol mono distearates mixed with fatty alcohol sulfates	TEGINACID	Goldschmidt
Glycerol mono distearates mixed with nonionics	TEGINACID	Goldschmidt
Glycerol monoisostearate	EMEREST	Quantum
Glycerol mono-oleate	ALKAMULS	Alkaril
Glycerol monooleate	EMEREST	Quantum
Glycerol monooleate	ESTOL	Unichema
Glycerol monooleate	DVR-EM	Van den Bergh
Glycerol monostearate	ALKAMULS	Alkaril
Glycerol monostearate	TEGIN	Goldschmidt
Glycerol monostearate*	EMEREST	Quantum

Chemical Name	Trade Name	Supplier
Glycerol monostearate	ESTOL	Unichema
Glyceryl monostearates from hydrogenated soybean oil	MYVAPLEX	Eastman
Glycerol triacetate (triacetin)	ESTOL	Unichema
Glycerol tricaprylate/-caprate	ESTOL	Unichema
Glycerol trioleate	ALKAMULS	Alkaril
Glycerol trioleate	EMEREST	Quantum
Glycerol trioleate	ESTOL	Unichema
Glyceryl dilaurate	GDL	Stepan
Glyceryl distearate	GDS	Stepan
Glyceryl laurate	GML	Stepan
Glyceryl oleate	CYCLOCHEM	Cyclo
Glyceryl oleate	DEHYMULS	Henkel
Glyceryl oleate*	MAZOL	Mazer
Glyceryl oleate	GMO	Stepan
Glyceryl oleate (and) propylene glycol	EXTAN	Lanaetex
Glyceryl stearate	DERMALCARE	Alcolac
Glyceryl stearate	CYCLOCHEM	Cyclo
Glyceryl stearate*	LIPO	Lipo
Glyceryl stearate*	MAZOL	Mazer
Glyceryl stearate	GMS	Stepan
Glyceryl stearate (and) PEG-stearate	CYCLOCHEM	Cyclo
Glyceryl stearate (and) PEG-stearate*	LIPOMULSE	Lipo
Glyceryl stearate (and) PEG-stearate*	MAZOL	Mazer
Glyceryl stearate (and) PEG-stearate		Stepan
Glycol distearate	CYCLOCHEM	Cyclo
Glycol distearate*	LIPO	Lipo
Glycol distearate*	MAPEG	Mazer
Glycol distearate	EGDS	Stepan
Glycol stearate	DERMALCARE	Alcolac
Glycol stearate (and) stearamide AMP	EGAS	Stepan
Guar gum		Gumix
Gum arabic		Gumix
Gum tragacanth		Gumix
Higher fatty alcohols and ethoxylates	CYCLOCHEM	Cyclo
Hydrogenated coconut acid	HYSTRENE, INDUSTRENE	Humko
Hydrogenated cottonseed oil	MYVEROL	Eastman
Hydrogenated palm oil	MYVEROL, MYVATEX	Eastman
Hydrogenated rapeseed oil, hydrogenated castor oil	MYVATEX	Eastman
Hydrogenated soybean oil	MYVEROL, MYVATEX	Eastman
Hydrogenated tallow amide*	ARMID	Akzo Chemical
Hydrogenated vegetable oil	MYVEROL	Eastman
Hydroxyethyl imidazoline-coconut	ALKAZINE	Alkaril
Hydroxyethyl imidazoline-oleic	ALKAZINE	Alkaril
Hydroxyethyl imidazoline-tall oil	ALKAZINE	Alkaril
Hydroxylated lanolin	OHLAN	Amerchol
Hydroxy stearamidopropyl trimonium chloride	SURFACTOL	CasChem

Chemical Name	Trade Name	Supplier
Hydroxy stearamidopropyl trimonium methyl sulfate	SURFACTOL	CasChem
Isobutyl oleate	ESTOL	Unichema
Isobutyl stearate	ESTOL	Unichema
Isooctyl oleate	ESTOL	Unichema
Isooctyl stearate	ESTOL	Unichema
Isopropylamine DBS	ALKASURF	Alkaril
Isopropyl myristate*	CYCLOCHEM	Cyclo
Isopropyl myristate	ESTOL	Unichema
Isopropyl oleate	ESTOL	Unichema
Isopropyl palmitate*	CYCLOCHEM	Cyclo
Isopropyl palmitate	ESTOL	Unichema
Isosteareth*	ETHLANA	Lanaetex
Isostearyl neopentanoate*	CYCLOCHEM	Cyclo
Isostearylpropyl dimethylamine*	MAZEEN	Mazer
Lactic acid ester of mono- and diglycerides	DURLAC	Van den Bergh
Lactic acid esters of monoglycerides	LACTODAN	Grindsted
Laneth-16*	LANAETEX	Lanaetex
Laneth-10 acetate*	LINSOL	Lanaetex
Laneth- (and) ceteth- (and) oleth- (and) steareth	SOLULAN	Amerchol
Lanolin*		Lanaetex
Lanolin alcohol*	ANATOL	Lanaetex
Lanolin (and) PEG-8-stearate*	LANOLA	Lanaetex
Lard	MYVEROL	Eastman
Lauramide DEA*	MAZAMIDE	Mazer
Lauramide DEA (and) propylene glycol	EMID	Quantum
Lauramidopropyl betaine*	MAFO	Mazer
Lauramine oxide*	MAZOX	Mazer
Lauramine oxide	AMMONYX	Stepan
Laureth-*	LANYCOL	Lanaetex
Laureth-*	LIPOCOL	Lipo
Laureth-*	MACOL	Mazer
Lauric acid*	NEO-FAT	Akzo Chemical
Lauric acid	HYSTRENE	Humko
Lauryl dimethyl amine oxide	ALKAMOX	Alkaril
Lauryl lactate	DERMALCARE	Alcolac
Lauryl lactate	CYCLOCHEM	Cyclo
Lecithin	ACTIFLO, CENTROLENE, CENTROLEX, CENTROMIX, CENTROPHASE	Central Soya
Lecithin	CLEARATE	Cleary
Lecithin	SSH	Eastman
Lignin-amine	INDULIN	Westvaco
Linear alcohol ethoxylate	DESONIC	DeSoto
Linear primary alcohols	NEODOL	Shell Chemical
Linoleamide DEA*	MAZAMIDE	Mazer
Locust bean gum	GELLOID	FMC
Locust bean gum		Gumix
Methyl ester of tallow fatty acids	ESTOL	Unichema
Methyl gluceth-20 distearate	GLUCAM	Amerchol
Methyl glucose dioleate	GLUCATE	Amerchol
Methyl laurate	ESTOL	Unichema

Chemical Name Index

Chemical Name	Trade Name	Supplier
Methyl oleate	ESTOL	Unichema
Methyl palmitate	ESTOL	Unichema
Methyl stearate	ESTOL	Unichema
Microcrystalline wax	DEHYMULS	Henkel
Mixed base phosphate ester	ALKAPHOS	Alkaril
Mono and diester laurates, stearates and oleates	CYCLOCHEM	Cyclo
Mono- and diglycerides	CAPMUL	Capital City
Mono- and diglycerides	ATMOS, ATMUL	Humko
Mono- and diglycerides	DUR-EM, DUR-LO, ICE, TALLY	Van den Bergh
Mono-diglycerides	EMULDAN	Grindsted
Monoglycerides	MIGHTY SOFT, MONOSET, MYVATEX, MYVEROL, SSH, TEXTURE LITE	Eastman
Monoglycerides	DIMODAN	Grindsted
Monoglycerides	ALPHADIM, DOUBLE SOFT, STARPLEX	Patco
Myreth*	LIPOCOL	Lipo
Myristamine oxide	MAZOX	Mazer
Myristamine oxide	AMMONYX	Stepan
Myristic acid	HYSTRENE	Humko
Myristyl lactate*	CYCLOCHEM	Cyclo
Myristyl myristate*	DERMALCARE	Alcolac
Myristyl myristate	CYCLOCHEM	Cyclo
Myristyl stearate*	DERMALCARE	Alcolac
Myristyl stearate*	CYCLOCHEM	Cyclo
Neopentyl glycol dioleate	ESTOL	Unichema
Nonionic: alcohol/ethylene oxide adducts	MERPOL	DuPont
Nonionic blend	EMULGATOR	Goldschmidt
Nonionic blend	T-MULZ, TOXIMUL	Harcros Chemicals
Nonionic blend	MICRO-STEP	Stepan
Nonionic ether	TRYLON	Quantum
Nonionic ethoxylate	ALKAMULS, ALKASURF	Alkaril
Nonionic fluorosurfactants	ZONYL	DuPont
Nonionic pigment dispersant	SYN FAC	Milliken
Nonionic wetting agent	TRYLON	Quantum
Nonoxynol-*	MACOL	Mazer
Nonyl nonoxynol-*	MACOL	Mazer
Nonyl phenol ethoxy (10)	WITCONOL NP-100	Witco Chemical
Nonyl phenol ethoxylate	ALKASURF	Alkaril
Nonyl phenol ethoxy phosphate ester	EMPHOS	Witco Chemical
Nonyl phenol ethoxy/propoxy	WITCONOL	Witco Chemical
Nonyl phenol ethoxy sulfate	WITCOLATE	Witco Chemical
Nutricol konjac flour with additives	NUTRICOL	FMC
Octaglycol monoleate	SANTONE	Van den Bergh
Octoxynol-*	MACOL	Mazer
Octyl phenol ethoxylate	ALKASURF	Alkaril
Octyl phenol ethoxylate	DESONIC	DeSoto
Oleamide*	ARMID	Akzo Chemical
Oleamide DEA*	MAZAMIDE	Mazer
Oleamidopropyl dimethylamine ethonium ethosulfate*	NAETEX	Lanaetex

Chemical Name	Trade Name	Supplier
Oleamidopropyl dimethylamine glycolate*	NAETEX	Lanaetex
Oleamine oxide*	MAZOX	Mazer
Oleic diethanolamide	EMID	Quantum
Oleth-	EUMULGIN	Henkel
Oleth-*	ETHOXOL	Lanaetex
Oleth-*	LIPOCOL	Lipo
Oleth-*	MACOL	Mazer
Oleyl alcohol ethoxylate	ALKASURF	Alkaril
Oleyl betaine*	MAFO	Mazer
Oleyl dimethyl amine oxide	ALKAMOX	Alkaril
Palmitaine oxide*	MAZOX	Mazer
Palmitic acid*	NEO-FAT	Akzo Chemical
Palmitic acid	HYSTRENE, INDUSTRENE	Humko
Palm oil	MYVEROL	Eastman
C12-15 Pareth-6-carboxylic acid*	SANDOPAN	Sandoz
Pentaerythritol cocoate	DEHYMULS	Henkel
Pentaerythritol tetraoleate*	CYCLOCHEM	Cyclo
Pentaerythritol tetraoleate*	MAZOL	Mazer
Pentaerythritol tetraoleate	ESTOL	Unichema
Pentaerythritol tetrasearate	CYCLOCHEM	Cyclo
PEG- castor oil	SURFACTOL	CasChem
PEG- castor oil*	MAPEG	Mazer
PEG- cocamide	AMIDOX	Stepan
PEG- cocamine*	ETHOMEEN	Akzo Chemical
PEG- cocamine*	MAZEEN	Mazer
PEG- cocoate*	ETHOFAT	Akzo Chemical
PEG- cocomonium chloride*	ETHOQUAD	Akzo Chemical
PEG- dilaurate	CYCLOCHEM	Alcolac
PEG- dilaurate*	LIPOPEG	Lipo
PEG- dilaurate*	MAPEGQ	Mazer
PEG- dilaurate	EMEREST	Quantum
PEG- dioleate*	LIPOPEG	Lipo
PEG- dioleate*	MAPEG	Mazer
PEG- dioleate	EMEREST	Quantum
PEG- distearate	CYCLOCHEM	Alcolac
PEG- distearate*	LIPOPEG	Lipo
PEG- distearate*	MAPEG	Mazer
PEG- distearate	EMEREST	Quantum
PEG- ditallate*	MAPEG	Mazer
PEG- glyceryl cocoate*	MAZOL	Mazer
PEG- glyceryl stearate	CUTINA	Henkel
PEG- glyceryl stearate*	MAZOL	Mazer
PEG- hydrogenated castor oil	CREMOPHOR	BASF
PEG- hydrogenated castor oil*	MAPEG	Mazer
PEG- hydrogenated lanolin*	LIPOLAN	Lipo
PEG- lanolin	SOLULAN	Amerchol
PEG- lanolin*	LANOBASE, LANTOX	Lanaetex
PEG- lauramide	AMIDOX	Stepan
PEG- laurate*	CYCLOCHEM	Alcolac
PEG- laurate*	LIPO, LIPOPEG	Lipo
PEG- laurate*	MAPEG	Mazer
PEG- laurate and dilaurate	PEG	Stepan
PEG- monoisostearate	TRYDET	Quantum
PEG- monolaurate	EMEREST	Quantum
PEG- monooleate	EMEREST	Quantum
PEG- monopelargonate*	EMEREST	Quantum
PEG- monostearate	EMEREST	Quantum

Chemical Name	Trade Name	Supplier
PEG- oleamide*	ETHOMID	Akzo Chemical
PEG- oleamine*	ETHOMEEN	Akzo Chemical
PEG- oleamonium chloride*	ETHOQUAD	Akzo Chemical
PEG- oleate*	ETHOFAT	Akzo Chemical
PEG- oleate	CYCLOCHEM	Alcolac
PEG- oleate*	LIPOPEG	Lipo
PEG- oleate*	MAPEG	Mazer
PEG- oleate and dioleate	PEG	Stepan
PEG- sesquioleate	EMEREST	Quantum
PEG- sorbitan laurate*	LIPOSORB	Lipo
PEG- sorbitan laurate*	T-MAZ	Mazer
PEG- soyamine*	ETHOMEEN	Akzo Chemical
PEG- soyamine*	MAZEEN	Mazer
PEG- soya sterol	GENEROL	Henkel
PEG- stearamine*	ETHOMEEN	Akzo Chemical
PEG- stearamonium chloride*	ETHOQUAD	Akzo Chemical
PEG- stearate*	CYCLOCHEM, DERMALCARE	Alcolac
PEG- stearate*	LANOXIDE	Lanaetex
PEG- stearate*	LIPOPEG	Lipo
PEG- stearate*	MAPEG	Mazer
PEG- stearate		Stepan
PEG- tallate*	MAPEG	Mazer
PEG- tallow amide*	ETHOMID	Akzo Chemical
PEG- tallow amine*	ETHOMEEN	Akzo Chemical
PEG- tallow amine*	MAZEEN	Mazer
PEG- trihydroxystearin	NATURECHEM	CasChem
Petrolatum*	LANAETEX	Lanaetex
Phosphate ester	DESOPHOS	DeSoto
Phosphate ester	TRYFAC	Quantum
Phosphate ester, hydrophobe: aliphatic	JORDAPHOS	Jordan
Phosphate ester, hydrophobe: aminic	JORDAPHOS	Jordan
Phosphate ester, hydrophobe: aromatic	JORDAPHOS	Jordan
Phosphate ester, hydrophobe: blend	JORDAPHOS	Jordan
Phospholipids	CENTROL, CENTROPHIL	Central Soya
POE- castor oil	TRYLOX	Quantum
POE- cocamide MEA*	MAZAMIDE	Mazer
POE- cocoamine	TRYMEEN	Quantum
POE- decylalcohol	TRYCOL	Quantum
POE- dinonylphenol	TRYCOL	Quantum
POE- hydrogenated castor oil	TRYLOX	Quantum
POE- lauramide DEA*	MAZAMIDE	Mazer
POE- lauryl alcohol	TRYCOL	Quantum
POE- nonylphenol	TRYCOL	Quantum
POE- octylphenol	TRYCOL	Quantum
POE- oleic acid	TRYPET	Quantum
POE- oleyl alcohol	TRYCOL	Quantum
POE- oleyl amine	TRYMEEN	Quantum
POE- sorbitan monolaurate	EMSORB	Quantum
POE- sorbitan monooleate	EMSORB	Quantum
POE- sorbitan monostearate	EMSORB	Quantum
POE- sorbitan trioleate	EMSORB	Quantum
POE- sorbitan tristearate	EMSORB	Quantum
POE- sorbitol	TRYLOX	Quantum
POE- sorbitol hexaoleate	TRYLOX	Quantum
POE- stearic acid	EMEREST, TRYDET	Quantum

Chemical Name	Trade Name	Supplier
POE- stearyl alcohol	TRYCOL	Quantum
POE- stearyl amine	TRYMEEN	Quantum
POE- tallow amine	TRYMEEN	Quantum
POE- tallow propylene diamine	TRYMEEN	Quantum
POE- tridecyl alcohol	TRYCOL	Quantum
Poligeenans		FMC
Poly basic fatty acids	INDULIN	Westvaco
Polyester surfactant	HYPERMER	ICI Specialty Chemicals
Polyethylene glycol esters		Hercules
Polyethylene glycol (400) mono- and dioleate	DURPREE	Van den Bergh
Polyethylene glycol 400 stearate	CREMOPHOR	BASF
Polygalactoside		Gumix
Polyglycerate 60	DURFAX	Van den Bergh
Polyglycerol esters of fatty acids	ICE	Van den Bergh
Polyglyceryl-3 oleate*	MAZOL	Mazer
Polyglyceryl-10 tetraoleate*	MAZOL	Mazer
Polyoxyaryl ether	SYN FAC	Milliken
Polyoxyethylene glycerol mono-laurate	TAGAT	Goldschmidt
Polyoxyethylene glycerol mono-oleate	TAGAT	Goldschmidt
Polyoxyethylene glycerol mono-stearate	TAGAT	Goldschmidt
Polyoxyethylene sorbitan mono-laurate	ALKAMULS	Alkaril
Polyoxyethylene sorbitan mono-oleate	ALKAMULS	Alkaril
Polyoxyethylene sorbitan mono-oleate	DURFAX	Van den Bergh
Polyoxyethylene sorbitan mono-stearate	ALKAMULS	Alkaril
Polyoxyethylene sorbitan mono-stearate	DURFAX	Van den Bergh
Polyoxyethylene sorbitan tri-oleate	ALKAMULS	Alkaril
Polyoxyethylene sorbitan tri-stearate	ALKAMULS	Alkaril
Polysorbate-	SOLULAN	Amerchol
Polysorbate	MSPS	Eastman
Polysorbate-*	LAXAN	Lanaetex
Polysorbate-*	LIPOSORB	Lipo
Polysorbate	LONZEST	Lonza
Polysorbate	T-MAZ	Mazer
Polysorbate	DURFAX, ICE	Van den Bergh
Potassium dihydroxyethyl cocamine oxide phosphate	MAZOX	Mazer
Potassium salt of phosphated alkoxylated aryl phenol	SYN FAC	Milliken
PPG- ceteareth-	EMULGIN	Henkel
PPG- Lanolin alcohol ether	SOLULAN	Amerchol
PPG- Lanolin ether	SOLULAN	Amerchol
Propylene glycol	DEHYMULS	Henkel
Propylene glycol dicaprylate/-caprate	ESTOL	Unichema
Propylene glycol dioleate	ESTOL	Unichema
Propylene glycol esters of fatty acids	PROMODAN	Grindsted
Propylene glycol laurate		Stepan
Propylene glycol (and) oleth-44*	ETHOXOL	Lanaetex
Propylene glycol mono- and di-esters	DURPRO	Van den Bergh
Propylene glycol monoesters	MYVATEX, MYVEROL, TEXTURE LITE	Eastman

Chemical Name	Trade Name	Supplier
Propylene glycol monostearate	TEGIN	Goldschmidt
Propylene glycol stearate*	CYCLOCHEM	Cyclo
Propylene glycol stearate*	MAZOL	Mazer
Propylene glycol stearate		Stepan
Quaternium-18	KEMAMINE	Humko
Rapeseed oil	MONOSET, MYVEROL	Eastman
Ricinoleamido-propyl tri-monium chloride	SURFACTOL	CasChem
Seaweed flour	GELCARIN	FMC
Secondary sodium alkane sulphonates		Mobay
Silicon dioxide	TEXTURE LITE	Eastman
Sodium alkyl ether sulfate	DUPONOL	DuPont
Sodium bistridecyl sulfo-succinate	AEROSOL	American Cyanamid
Sodium branched dodecylbenzene sulfonate	WITCONATE	Witco Chemical
Sodium cetearyl sulfate	Lanette	Henkel
Sodium ceteth carboxylate	SANDOPAN	Sandoz
Sodium diamyl sulfosuccinate	AEROSOL	American Cyanamid
Sodium dicyclohexyl sulfo-succinate	AEROSOL	American Cyanamid
Sodium dihexyl sulfosuccinate	AEROSOL	American Cyanamid
Sodium diisobutyl sulfonate	AEROSOL	American Cyanamid
Sodium diisopropyl naphthalene sulfonate	AEROSOL	American Cyanamid
Sodium dioctyl sulfosuccinate	AEROSOL	American Cyanamid
Sodium C16-20 ethoxylate carboxylate*	SANDOPAN	Sandoz
Sodium 2-ethylhexyl sulfate	SIPEX	Alcolac
Sodium isodecyl sulfate	SIPEX	Alcolac
Sodium laureth-13-carboxylate	SANDOPAN	Sandoz
Sodium lauryl sulfate	DUPONOL	DuPont
Sodium lauryl sulfate	WITCOLATE	Witco Chemical
Sodium lauryl/oleyl sulfate	DUPONOL	DuPont
Sodium lauryl/propoxy sulfo-succinate	EMCOL	Witco Chemical
Sodium neutralized condensed naphthalene sulfuric acid	AEROSOL	American Cyanamid
Sodium octyl/decyl sulfate	DUPONOL	DuPont
Sodium octyl sulfate	DUPONOL	DuPont
Sodium C14-16 olefin sulfonate	WITCONATE	Witco Chemical
Sodium oleyl/lauryl sulfate	DUPONOL	DuPont
Sodium C12-15 pareth carboxylate	SANDOPAN	Sandoz
Sodium C12-15 pareth sulfonate*	AVANEL	Mazer
Sodium petroleum sulfonate	PETRONATE	Sonneborn
Sodium salt of sulfated oleyl acetate	AVITEX	DuPont
Sodium stearoyl lactylate	TEXTURE LITE	Eastman
Sodium stearoyl lactylate	ARTODAN	Grindsted
Sodium stearoyl lactylate	EMPLEX	Patco
Sodium tridecyl ether sulfate	SIPEX	Alcolac
Sodium trideceth carboxylate*	SANDOPAN	Sandoz
Sodium tridecyl sulfate	SIPEX	Alcolac
Sorbitan diisostearate	EMSORB	Quantum
Sorbitan esters		Humko

Chemical Name	Trade Name	Supplier
Sorbitan esters of fatty acids	FAMODAN	Grindsted
Sorbitan laurate*	EXTAN	Lanaetex
Sorbitan laurate*	LIPOSORB	Lipo
Sorbitan laurate	LONZEST	Lonza
Sorbitan laurate*	S-MAZ	Mazer
Sorbitan monoisostearate	EMSORB	Quantum
Sorbitan monolaurate	ALKAMULS	Alkaril
Sorbitan monolaurate	EMSORB	Quantum
Sorbitan monooleate	ALKAMULS	Alkaril
Sorbitan monooleate*	EMSORB	Quantum
Sorbitan monopalmitate	EMSORB	Quantum
Sorbitan monostearate	ALKAMULS	Alkaril
Sorbitan monostearate	EMSORB	Quantum
Sorbitan monostearate	DURTAN	Van den Bergh
Sorbitan oleate*	EXTAN	Lanaetex
Sorbitan oleate	LIPOSORB	Lipo
Sorbitan oleate	LONZEST	Lonza
Sorbitan oleate	S-MAZ	Mazer
Sorbitan palmitate*	EXTAN	Lanaetex
Sorbitan palmitate	LIPOSORB	Lipo
Sorbitan palmitate	LONZEST	Lonza
Sorbitan palmitate	S-MAZ	Mazer
Sorbitan sesquioleate	DEHYMULS	Henkel
Sorbitan sesquioleate*	EXTAN	Lanaetex
Sorbitan sesquioleate	LIPOSORB	Lipo
Sorbitan sesquioleate	S-MAZ	Mazer
Sorbitan sesquioleate	EMSORB	Quantum
Sorbitan stearate	EXTAN	Lanaetex
Sorbitan stearate*	LIPOSORB	Lipo
Sorbitan stearate	LONZEST	Lonza
Sorbitan stearate*	S-MAZ	Mazer
Sorbitan trioleate	ALKAMULS	Alkaril
Sorbitan trioleate*	LIPOSORB	Lipo
Sorbitan trioleate	LONZEST	Lonza
Sorbitan trioleate	S-MAZ	Mazer
Sorbitan trioleate	EMSORB	Quantum
Sorbitan tristearate	ALKAMULS	Alkaril
Sorbitan tristearate	LIPOSORB	Lipo
Sorbitan tristearate	LONZEST	Lonza
Sorbitan tristearate	S-MAZ	Mazer
Soya sterol	GENEROL	Henkel
Soybean oil	CENTROL	Central Soya
Soybean organic phosphatides	KELECIN	NL Chemicals
Stearalkonium chloride	AMMONYX	Stepan
Stearamide*	ARMID	Akzo Chemical
Stearamide DEA*	LIPAMIDE	Lipo
Stearamide MEA*	MAZAMIDE	Mazer
Stearamidoethyl diethylamine	CHEMICAL BASE	Sandoz
Stearamidoethyl ethanolamine	CHEMICAL BASE	Sandoz
Stearamidopropyl diethylamine*	MAZEEN	Mazer
Stearamidopropyl dimethylamine lactate*	MAZEEN	Mazer
Stearamine oxide*	MAZOX	Mazer
Stearamine oxide	AMMONYX	Stepan
Steareth*	LIPOCOL	Lipo
Steareth*	MACOL	Mazer
Stearic acid*	NEO-FAT	Akzo Chemical
Stearic acid	HYSTRENE, INDUSTRENE	Humko
Stearyl alcohol	PROMULGEN	Amerchol
Stearyl alcohol*	LIPOWAX	Lipo

Chemical Name	Trade Name	Supplier
Stearyl alcohol ethoxylate	ALKASURF	Alkaril
Stearyl citrate	DEHYMULS	Henkel
Stearyl stearate*	CYCLOCHEM	Cyclo
Succinylated monoglycerides	DO CONTROL, MYVEROL	Eastman
Sulfated castor oil	ACTRASOL	Climax Performance
Sulfated ricinoleic acid	ACTRASOL	Climax Performance
Sulfated soya ester	ACTRASOL	Climax Performance
Sulfonate, nonionic blend	MICRO-STEP, TOXIMUL	Stepan
Sulphonates, aliphatic	ALKANOL, PETROWET	DuPont
Sulphonates, alkylaryl	ALKANOL	DuPont
Sunflower oil	MYVEROL	Eastman
Tall oil fatty amide	EMULAMID	Mayco
Tallow alcohol ethoxylate	ALKASURF	Alkaril
Tallow alkonium chloride	KEMAMINE	Humko
Tallow amine, POE	TOXIMUL	Stepan
Tallow trimonium chloride	KEMAMINE	Humko
Tetrahydroxypropyl ethylenediamine*	MAZEEN	Mazer
Tetrasodium N-(1,2-dicarboxyethyl)-N-octadecyl sulfosuccinamate	AEROSOL	American Cyanamid
Triacetin TEGDA blend	ESTOL	Unichema
Trideceth-*	LIPOCOL	Lipo
Trideceth-*	MACOL	Mazer
Trideceth carboxylic acid*	SANDOPAN	Sandoz
Triethanolamine DBS	ALKASURF	Alkaril
Triethylene glycol diacetate (TEGDA)	ESTOL	Unichema
Triglycerides	KELECIN	NL Chemicals
Triglycerol monostearate	SANTONE	Van den Bergh
Triglyceryl diisostearate	EMEREST	Quantum
Triisostearin	DERMALCARE	Alcolac
Triisostearin*	CYCLOCHEM	Cyclo
Trilaurin*	CYCLOCHEM	Cyclo
Trimethylol-propane trioleate	ESTOL	Unichema
Triolein*	CYCLOCHEM	Cyclo
Tris(hydroxymethyl)aminomethane	TRIS AMINO	Angus

TRADE NAME INDEX

Trade Name	Supplier
ABEX	Alcolac
ACTIFLO	Central Soya
ACTRABASE	Climax Performance Materials
ACTRASOL	Climax Performance Materials
AEROSOL	American Cyanamid Co.
ALKAMIDE	Alkaril Chemicals Ltd.
ALKAMIDOX	Alkaril Chemicals Ltd.
ALKAMOX	Alkaril Chemicals Ltd.
ALKAMULS	Alkaril Chemicals Ltd.
ALKANOL	DuPont Co.
ALKAPHOS	Alkaril Chemicals Ltd.
ALKASURF	Alkaril Chemicals Ltd.
ALKAZINE	Alkaril Chemicals Ltd.
ALPHADIM	Patco Products
AMIDAN	Grindsted Products, Inc.
AMIDOX	Stepan Co.
AMMONYX	Stepan Co.
ANATOL	Lanaetex Products, Inc.
ARMID	Akzo Chemicals, Inc.
ARTODAN	Grindsted Products, Inc.
ATLAS	ICI Specialty Chemicals
ATLOX	ICI Specialty Chemicals
ATMOS	Humko Chemical Division
ATMUL	Humko Chemical Division
AVITEX	DuPont
CAPMUL	Capital City Products Co.
CASUL	Harcros Chemicals Inc.
CATIMULS	ScanRoad Inc.
CENTROL	Central Soya
CENTROLENE	Central Soya
CENTROLEX	Central Soya
CENTROMIX	Central Soya
CENTROPHASE	Central Soya
CENTROPHIL	Central Soya
CETODAN	Grindsted Products, Inc.
CLEARATE	W.A. Cleary Products, Inc.
CREMOPHOR	BASF Corp.
CUTINA	Henkel Corp.
CYCLOCHEM	Alcolac
CYCLOCHEM	Cyclo Chemicals Corp.
DEHYMULS	Henkel Corp.
DERMALCARE	Alcolac
DESOMEEN	DeSoto, Inc.
DESONIC	DeSoto, Inc.
DESOPHOS	DeSoto, Inc.
DESOTAN	DeSoto, Inc.
DIMODAN	Grindsted Products, Inc.
DO CONTROL	Eastman Chemical Products, Inc.
DOUBLE SOFT	Patco Products
DUPONOL	DuPont Co.
DUR-EM	Van den Bergh Food Ingredients Group
DURFAX	Van den Bergh Food Ingredients Group
DURLAC	Van den Bergh Food Ingredients Group
DUR-LO	Van den Bergh Food Ingredients Group
DURPREE	Van den Bergh Food Ingredients Group
DURPRO	Van den Bergh Food Ingredients Group
DURTAN	Van den Bergh Food Ingredients Group
EMCOL	Witco Corp.
EMEREST	Quantum Chemical Corp.
EMERY	Quantum Chemical Corp.
EMID	Quantum Chemical Corp.
EMPHOS	Witco Corp.

Trade Name	Supplier
EMPLEX	Patco Products
EMSORB	Quantum Chemical Corp.
EMULAMID	Mayco Oil & Chemical Co.
EMULDAN	Grindsted Products, Inc.
EMULGATOR	Goldschmidt Chemical Corp.
ESTOL	Unichema Chemicals, Inc.
ETHLANA	Lanaetex Products, Inc.
ETHOFAT	Akzo Chemicals, Inc.
ETHOMEEN	Akzo Chemicals, Inc.
ETHOMID	Akzo Chemicals, Inc.
ETHOQUAD	Akzo Chemicals, Inc.
ETHOXOL	Lanaetex Products, Inc.
ETHOXYCHOL	Lanaetex Products, Inc.
EUMULGIN	Henkel Corp.
EXTAN	Lanaetex Products, Inc.
FAMODAN	Grindsted Products, Inc.
GELCARIN	FMC Corp.
GELLOID	FMC Corp.
GENEROL	Henkel Corp.
GLUCAM	Amerchol Corp.
GLUCATE	Amerchol Corp.
HYPERMER	ICI Specialty Chemicals
HYSTRENE	Humko Chemical Division
ICE	Van den Bergh Food Ingredients Group
INDULIN	Westvaco Chemicals
INDUSTRENE	Humko Chemical Division
JORDAPHOS	Jordan Chemical Co.
KEMAMINE	Humko Chemical Division
LACTARIN	FMC Corp.
LACTODAN	Grindsted Products, Inc.
LANAETEX	Lanaetex Products, Inc.
LANETTE	Henkel Corp.
LANOBASE	Lanaetex Products, Inc.
LANOLA	Lanaetex Products, Inc.
LANOXIDE	Lanaetex Products, Inc.
LANTOX	Lanaetex Products, Inc.
LANYCOL	Lanaetex Products, Inc.
LAXAN	Lanaetex Products, Inc.
LINSOL	Lanaetex Products, Inc.
LIPAMIDE	Lipo Chemicals Inc.
LIPO	Lipo Chemicals Inc.
LIPOCOL	Lipo Chemicals Inc.
LIPODAN	Grindsted Products, Inc.
LIPOLAN	Lipo Chemicals Inc.
LIPOMULSE	Lipo Chemicals Inc.
LIPOPEG	Lipo Chemicals Inc.
LIPOSORB	Lipo Chemicals Inc.
LIPOWAX	Lipo Chemicals Inc.
LIQUID LITE	Eastman Chemical Products, Inc.
LONZEST	Lonza Inc.
MAYSOL	Mayco Oil & Chemical Co.
MERPOL	DuPont Co.
MICRO-STEP	Stepan Co.
MIGHTY SOFT	Eastman Chemical Products, Inc.
MONOSET	Eastman Chemical Products, Inc.
MYVASET	Eastman Chemical Products, Inc.
MYVAPLEX	Eastman Chemical Products, Inc.
MYVATEM	Eastman Chemical Products, Inc.
MYVATEX	Eastman Chemical Products, Inc.
MYVEROL	Eastman Chemical Products, Inc.
NAETEX	Lanaetex Products, Inc.
NATURECHEM	CasChem, Inc.
NEODOL	Shell Chemical Co.

Trade Name	Supplier
NEO-FAT	Akzo Chemicals, Inc.
NIPOL	Stepan Co.
NUTRICOL	FMC Corp.
OHLAN	Amerchol Corp.
PANODAN	Grindsted Products, Inc.
PETRONATE	Sonneborn Division
PETROWET	DuPont Co.
POLIGEENAN	FMC Corp.
PRICERINE	Unichema Chemicals, Inc.
PROMODAN	Grindsted Products, Inc.
PROMULGEN	Amerchol Corp.
SANDOPAN	Sandoz Chemicals Corp.
SANTONE	Van den Bergh Food Ingredients Group
SEAGEL	FMC Corp.
SEAKEM	FMC Corp.
SEASPEN	FMC Corp.
SIPEX	Alcolac
SOLULAN	Amerchol Corp.
STARPLEX	Patco Products
STEPFAC	Stepan Co.
SURFACTOL	CasChem, Inc.
SYN FAC	Milliken Chemical
TAGAT	Goldschmidt Chemical Corp.
TALLY	Van den Bergh Food Ingredients Group
TANDEM	Humko Chemical Division
TEGIN	Goldschmidt Chemical Corp.
TEGINACID	Goldschmidt Chemical Corp.
TEXTURE LITE	Eastman Chemical Products, Inc.
T-MULZ	Harcros Chemicals Inc.
TOXIMUL	Stepan Co.
TRIODAN	Grindsted Products, Inc.
TRYCOL	Quantum Chemical Corp.
TRYDET	Quantum Chemical Corp.
TRYFAC	Quantum Chemical Corp.
TRYLON	Quantum Chemical Corp.
TRYLOX	Quantum Chemical Corp.
TRYMEEN	Quantum Chemical Corp.
TWEEN	ICI Specialty Chemicals
VERV	Patco Products
VISCARIN	FMC Corp.
WITCOLATE	Witco Corp.
WITCONATE	Witco Corp.
WITCONOL	Witco Corp.
ZELEC	DuPont Co.
ZONYL	DuPont Co.

Other Noyes Publications

CHEMICAL GUIDE TO THE UNITED STATES 1989/1990
Eighth Edition

Edited by
D.J. De Renzo

This Eighth Edition of the *Chemical Guide to the United States* describes more than **220 of the largest U.S. chemical companies.** The firms are organized alphabetically by parent company, with all chemical manufacturing divisions, subsidiaries, and affiliates included under the parent firm's heading. Including parent companies, divisions, subsidiaries, affiliates, and plant locations, the book contains **more than 2000 entries.**

The U.S. chemical industry set sales records during 1988 with an estimated $240 billion in chemical sales (as defined by the Chemical Manufacturers Association). The chemical industry contributed to the U.S. trade balance in a positive manner. Exports were about $32 billion, while imports were about $21 billion. Capital expenditures for new plants and equipment rose to a record level—about $18.5 billion; and research and development spending was about $10.6 billion, another record. The chemical industry currently employs well over a million people, thus accounting for a sizeable portion of the gross national product.

The information presented for each parent company includes, as available:

- Name and Address of Headquarters
- Ownership
- Sales Figures
- Principal Executives
- Domestic Subsidiaries and Affiliates
- Plant Locations—Addresses and Phone Numbers
- Names of Plant Managers
- Products Produced at Plant Locations

Also included in the book are two very useful indexes, an **Index of Parent Companies, Divisions, Subsidiaries and Affiliates** and a **Geographical (Zip Code) Index.**

The companies listed are those that have annual chemical sales greater than $10 million. Information is also included on privately held firms, joint ventures and others that do not publish annual reports. The companies described actually carry out chemical reactions in their plants. Not included are companies which primarily process chemicals physically rather than chemically. Pharmaceutical firms are included only if they also produce commercial chemicals and intermediates.

The reader will find this valuable Guide helpful in many ways:

- **as a valuable market research tool**
- **to find information quickly**
- **to increase sales to the chemical industry**
- **to search for potential acquisitions and divestitures**
- to know whom to contact
- as a useful employment guide
- **to pinpoint sales efforts to BIG BUYERS**
- **to organize sales efforts geographically**
- **to identify joint venture companies**
- **to obtain hard-to-get information on medium-sized and smaller firms**
- to gain a broad picture of the U.S. chemical industry
- to locate data on recent mergers and joint ventures
- to find out who owns whom

The **Geographical (Zip Code) Index** should be of great assistance in market research studies; and help your personnel locate all other chemical plants in whatever area they happen to be; as well as being helpful for "over-the-fence" studies.

ISBN 0-941459-01-2 (1989) 6"x9" 443 pages

Other Noyes Publications

ADVANCED CLEANING PRODUCT FORMULATIONS
Household, Industrial, Automotive

by

Ernest W. Flick

This book presents **more than 800 up-to-date advanced cleaning product formulations** for household, industrial and automotive applications. It is the result of information received from numerous industrial companies and other organizations. The data represent selections made at no cost to, nor influence from, the makers or distributors of these materials. Only the most recent formulas have been included. It is believed that all of the trademarked raw materials listed here are currently available.

The formulations in the book are divided as follows. Parenthetic numbers indicate the number of formulations in each chapter.

I. HOUSEHOLD/INDUSTRIAL CLEANERS

1. **Bathroom Cleaners (16)**
2. **Disinfectants (11)**
3. **Dishwashing Detergents (57)**
4. **Floor Cleaners and Wax Strippers (41)**
5. **General Purpose Cleaners (73)**
6. **Laundry Products (143)**
7. **Metal Cleaners (74)**
8. **Oven Cleaners (10)**
9. **Rinse Additives and Aids (97)**
10. **Rug, Carpet and Upholstery Cleaners and Shampoos (41)**
11. **Wall and Hard Surface Cleaners (52)**
12. **Window and Glass Cleaners (25)**
13. **Miscellaneous Cleaners (130)**

II. AUTOMOTIVE CLEANERS

14. **Car and Truck Washes (48)**
15. **Whitewall Tire Cleaners (8)**
16. **Miscellaneous Cleaners (22)**

Each formulation in the book lists the following information, as available, in the manufacturer's own words:

- Description of end use and most outstanding properties.
- The percent by weight or volume of each raw material included in the formula, rounded to a decimal figure.
- Key properties of the formula, which are the features that the source considers to be more outstanding than other formulations of the same type.
- The formula source, which is the company or organization that supplied the formula. The secondary source may be the originating company and/or the primary source's publication title, or both. A formula number is included, if applicable.

In addition to the sections listed above, there are two other sections which will be helpful to the reader:

III. A chemical trade name section where trade-named raw materials included in the book are listed with a brief chemical description and the supplier's name. The specifications which each raw material meets are included, if applicable.

IV. Main office addresses of the suppliers of trade-named raw materials.

ISBN 0-8155-1186-8 (1989)   6"x9"   372 pages

Other Noyes Publications

INDUSTRIAL SURFACTANTS

by

Ernest W. Flick

This book describes almost 3500 surfactants which are currently available for industrial use. The book will be of value to technical and managerial personnel involved in the specification and use of these products.

Industrial surfactants find uses in almost every industry, from asphalt manufacturing to carpet fibers, from petroleum production to leather and fur processing. Examples of the types of chemicals used as surfactants are fatty alcohol sulfates, alkylolamides, alkoxylates, sulfosuccinates, amines, quaternaries, phosphate esters, acid esters, block copolymers, betaines, imidazolines, alkyl sulfonates, etc. The market for these products has, and is expected to continue to have, a steady growth rate, of the order of 4 to 5%, and in some cases up to 10%, per year.

The data included represent selections from manufacturers' descriptions made at no cost to, nor influence from, the makers or distributors of the materials. Only the most recent information has been included. It is believed that all of the products listed are currently available, which will be of utmost interest to readers concerned with product discontinuances.

Products are presented by company, and the companies are listed alphabetically. Also included are a **Trade Name Index** and a list of **Suppliers' Addresses.**

The book lists the following product information, as available, in the manufacturer's own words: company name and product category, trade name and product numbers, plus a description of the product, as presented by the supplier.

The following companies are represented in the book:

Air Products and Chemicals
Akzo America, Inc.
Akzo Chemie America
Albright & Wilson, Inc.
Alcolac, Inc.
American Cyanamid Co.
American Hoechst Corp.
Angus Chemical Co.

BASF Corp.
Capital City Products Co.
Central Soya/
   Chemurgy Division
Continental Chemical Co.
Croda, Inc.
DeSoto, Inc.
Dow Chemical U.S.A.
DuPont Co.
Emery Industries, Inc.
Emulsion Systems, Inc.
Ethyl Corp
Exxon Chemical Co./
   Tomah Products
GAF Corp.
Goldschmidt Chemical Corp.
W.R. Grace & Co.
Henkel Corp.
ICI Americas, Inc.
Inolex Chemical Co.
Jordan Chemical Co.
Lonza, Inc.
Mazer Chemicals, Inc.
McIntyre Chemical Co.
Miranol, Inc.
Mona Industries, Inc.
Monomer-Polymer and Dajac
   Laboratories, Inc.
Monsanto Co.
Niacet Corp.
Olin Chemicals
Pilot Chemical Co.
Rohm and Haas Co.
Ruetgers-Nease Chemical Co.
Samson Chemical Co., Inc.
Sandoz Chemicals Corp.
Scher Chemicals, Inc.
Shell Chemical Co.
Sherex Chemical Co.
Stepan Co.
3M Corp.
Texaco Corp.
Arthur C. Trask Corp.
Union Carbide Corp.
R.T. Vanderbilt Co., Inc.
Vista Chemical Co.
Westvaco Chemicals Division
Witco Corp.

ISBN 0-8155-1173-6 (1988)   6"x9"   743 pages